W9-APX-266

The Age
of Empathy

The Age of Empathy

NATURE'S LESSONS

FOR A KINDER SOCIETY

Frans de Waal

With drawings by the author

HARMONY BOOKS • NEW YORK

All rights reserved.
Published in the United States by Harmony Books, an imprint of the Crown Publishing
Group, a division of Random House, Inc., New York.
www.crownpublishing.com

Harmony Books is a registered trademark and the Harmony Books colophon is a
trademark of Random House, Inc.

Library of Congress Cataloging-in-Publication Data is available upon request.

ISBN 978-0-307-40776-4

Printed in the United States of America

Design by Debbie Glasserman

10 9 8 7 6 5 4 3 2 1

First Edition

For Catherine, who makes me laugh

Contents

Preface

Greed is out, empathy is in.

The global financial crisis of 2008, together with the election of a new American president, has produced a seismic shift in society. Many have felt as if they were waking up from a bad dream about a big casino where the people's money had been gambled away, enriching a happy few without the slightest worry about the rest of us. This nightmare was set in motion a quarter century earlier by Reagan-Thatcher trickle-down economics and the soothing reassurance that markets are wonderful at self-regulation. No one believes this anymore.

American politics seems poised for a new epoch that stresses cooperation and social responsibility. The emphasis is on what unites a society, what makes it worth living in, rather than what material wealth we can extract from it. Empathy is the grand theme of our time, as reflected in the speeches of Barack Obama, such as when he told graduates at Northwestern University, in Chicago: "I think we should talk more about our empathy deficit.... It's only when you hitch your wagon to something larger than yourself that you will realize your true potential."

The message of *The Age of Empathy* is that human nature offers a giant helping hand in this endeavor. True, biology is usually called upon to justify a society based on selfish principles, but we should never forget that it has also produced the glue that holds communities together. This glue is the same for us as for many other animals. Being in tune with others, coordinating activities, and caring for those in need isn't restricted to our species. Human empathy has the backing of a long evolutionary history—which is the second meaning of "age" in this book's title.

Biology, Left and Right

What is government itself but the greatest of all reflections on human nature?

—JAMES MADISON, 1788

Are we our brothers' keepers? Should we be? Or would this role only interfere with why we are on earth, which according to economists is to consume and produce, and according to biologists is to survive and reproduce? That both views sound similar is logical given that they arose at around the same time, in the same place, during the English Industrial Revolution. Both follow a competition-is-good-for-you logic.

Slightly earlier and slightly to the north, in Scotland, the thinking was different. The father of economics, Adam Smith, understood as no other that the pursuit of self-interest needs to be tempered by "fellow feeling." He said so in *The Theory of Moral Sentiments*, a book not nearly as popular as his later work *The Wealth of Nations*. He famously opened his first book with:

How selfish soever man may be supposed, there are evidently
some principles in his nature, which interest him in the fortune of
others, and render their happiness necessary to him, though he
derives nothing from it except the pleasure of seeing it.

The French revolutionaries chanted of *fraternité*, Abraham Lincoln
appealed to the bonds of sympathy, and Theodore Roosevelt spoke
glowingly of fellow feeling as "the most important factor in producing a healthy political and social life." But if this is true, why is this
sentiment sometimes ridiculed as being, well, sentimental? A recent
example occurred after Hurricane Katrina struck Louisiana in 2005.
While the American people were transfixed by the unprecedented catastrophe, one cable news network saw fit to ask if the Constitution
actually provides for disaster relief. A guest on the show argued that
the misery of others is none of our business.

The day the levees broke, I happened to be driving down from Atlanta to Alabama to give a lecture at Auburn University. Except for a
few fallen trees, this part of Alabama had suffered little damage, but the
hotel was full of refugees: people had crammed the rooms with grand-
parents, children, dogs, and cats. I woke up at a zoo! Not the strangest
place for a biologist, perhaps, but it conveyed the size of the calamity.
And these people were the lucky ones. The morning newspaper at
my door screamed, "Why have we been left behind like animals?" a
quote from one of the people stuck for days without food and sanita-
tion in the Louisiana Superdome.

I took issue with this headline, not because I felt there was nothing to complain about, but because animals don't necessarily leave
one another behind. My lecture was on precisely this topic, on how
we have an "inner ape" that is not nearly as callous and nasty as adver-
tised, and how empathy comes naturally to our species. I wasn't claiming that it always finds expression, though. Thousands of people with
money and cars had fled New Orleans, leaving the sick, old, and poor
to fend for themselves. In some places dead bodies floated in the water,
where they were being eaten by alligators.

But immediately following the disaster there was also deep embarrassment in the nation about what had happened, and an incredible outpouring of support. Sympathy was not absent—it just was late in coming. Americans are a generous people, yet raised with the mistaken belief that the "invisible hand" of the free market—a metaphor introduced by the same Adam Smith—will take care of society's woes. The invisible hand, however, did nothing to prevent the appalling survival-of-the-fittest scenes in New Orleans.

The ugly secret of economic success is that it sometimes comes at the expense of public funding, thus creating a giant underclass that no one cares about. Katrina exposed the underbelly of American society. On my drive back to Atlanta, it occurred to me that this is the theme of our time: the common good. We tend to focus on wars, terror threats, globalization, and petty political scandals, yet the larger issue is how to combine a thriving economy with a humane society. It relates to health care, education, justice, and—as illustrated by Katrina—protection against nature. The levees in Louisiana had been criminally neglected. In the weeks following the flooding, the media were busy finger-pointing. Had the engineers been at fault? Had funds been diverted? Shouldn't the president have broken off his vacation? Where I come from, fingers belong in the dike—or at least that's how legend has it. In the Netherlands, much of which lies up to twenty feet below sea level, dikes are so sacred that politicians have literally no say over them: Water management is in the hands of engineers and local citizen boards that predate the nation itself.

Come to think of it, this also reflects a distrust of government, not so much *big* government but rather the short-sightedness of most politicians.

Evolutionary Spirit

How people organize their societies may not seem the sort of topic a biologist should worry about. I should be concerned with the ivory-billed woodpecker, the role of primates in the spread of AIDS or Ebola, the disappearance of tropical rain forests, or whether we evolved from the apes. Whereas the latter remains an issue for some, there has nevertheless been a dramatic shift in public opinion regarding the role of biology. The days are behind us when E. O. Wilson was showered with cold water after a lecture on the connection between animal and human behavior. Greater openness to parallels with animals makes life easier for the biologist, hence my decision to go to the next level and see if biology can shed light on human society. If this means wading right into political controversy, so be it; it's not as if biology is not already a part of it. Every debate about society and government makes huge assumptions about human nature, which are presented as if they come straight out of biology. But they almost never do.

Lovers of open competition, for example, often invoke evolution. The e-word even slipped into the infamous "greed speech" of Gordon Gekko, the ruthless corporate raider played by Michael Douglas in the 1987 movie *Wall Street*:

> The point is, ladies and gentleman, that "greed"—for lack of a better word—is good. Greed is right. Greed works. Greed clarifies, cuts through, and captures the essence of the evolutionary spirit.

The evolutionary spirit? Why are assumptions about biology always on the negative side? In the social sciences, human nature is typified by the old Hobbesian proverb *Homo homini lupus* ("Man is wolf to man"), a questionable statement about our own species based on false assumptions about another species. A biologist exploring the interaction between society and human nature really isn't doing anything new, therefore. The only difference is that instead of trying to justify a particular ideological framework, the biologist has an actual interest in the question of what human nature is and where it came from. Is the evolutionary spirit really all about greed, as Gekko claimed, or is there more to it?

Students of law, economics, and politics lack the tools to look at their own society with any objectivity. What are they going to compare it with? They rarely, if ever, consult the vast knowledge of human behavior accumulated in anthropology, psychology, biology, or neuroscience. The short answer derived from the latter disciplines is that we are group animals: highly cooperative, sensitive to injustice, sometimes warmongering, but mostly peace loving. A society that ignores these tendencies can't be optimal. True, we are also incentive-driven animals, focused on status, territory, and food security, so that any society that ignores those tendencies can't be optimal, either. There is both a social and a selfish side to our species. But since the latter is, at least in the West, the dominant assumption, my focus will be on the former: the role of empathy and social connectedness.

There is exciting new research about the origins of altruism and fairness in both ourselves and other animals. For example, if one gives two monkeys hugely different rewards for the same task, the one who gets the short end of the stick simply refuses to perform. In our own species, too, individuals reject income if they feel the distribution is unfair. Since *any* income should beat none at all, this means that both monkeys and people fail to follow the profit principle to the letter. By protesting against unfairness, their behavior supports both the claim that incentives matter and that there is a natural dislike of injustice.

Yet in some ways we seem to be moving ever closer to a society with no solidarity whatsoever, one in which a lot of people can expect the short end of the stick. To reconcile this trend with good old Christian values, such as care for the sick and poor, may seem hopeless. But one common strategy is to point the finger at the victims. If the poor can be blamed for being poor, everyone else is off the hook. Thus, a year after Katrina, Newt Gingrich, a prominent conservative politician, called for an investigation into "the failure of citizenship" of people who had been unsuccessful escaping from the hurricane.

Those who highlight individual freedom often regard collective interests as a romantic notion, something for sissies and communists. They prefer an every-man-for-himself logic. For example, instead of spending money on levees that protect an entire region, why not let everyone take care of their own safety? A new company in Florida is doing just that, renting out seats on private jets to fly people out of places threatened by hurricanes. This way, those who can afford it won't need to drive out at five miles per hour with the rest of the populace.

Every society has to deal with this me-first attitude. I see it play out every day. And here I am not referring to people, but to chimpanzees at the Yerkes National Primate Research Center, where I work. At our field station northeast of Atlanta, we house chimps in large outdoor corrals, sometimes providing them with shareable food, such as watermelons. Most of the apes want to be the first to put their hands on our food, because once they have it, it's rarely taken away by others. There actually exists respect of ownership, so that even the lowest-ranking female is allowed to keep her food by the most dominant male. Food possessors are often approached by others with an outstretched hand (a gesture that is also the universal way humans ask for a handout). The apes beg and whine, literally whimpering in the face of the other. If the possessor doesn't give in, beggars may throw a fit, screaming and rolling around as if the world is coming to an end.

My point is that there is both ownership and sharing. In the end, usually within twenty minutes, all of the chimpanzees in the group will have some food. Owners share with their best buddies and family, who in turn share with their best buddies and family. It is a rather peaceful scene even though there is also quite a bit of jostling for position. I still remember a camera crew filming a sharing session and the cameraman turning to me and saying, "I should show this to my kids. They could learn from it."

So, don't believe anyone who says that since nature is based on a struggle for life, we need to live like this as well. Many animals survive not by

Chimps beg for a share of food with the same palm-up gesture typical of our own species.

eliminating each other or keeping everything for themselves, but by cooperating and sharing. This applies most definitely to pack hunters, such as wolves or killer whales, but also to our closest relatives, the primates. In a study done at Taï National Park, in Ivory Coast, chimpanzees took care of group mates wounded by leopards; they licked their mates' blood, carefully removed dirt, and waved away flies that came near the wounds. They protected injured companions and slowed down during travel in order to accommodate them. All of this makes perfect sense, given that chimpanzees live in groups for a reason, the same way wolves and humans are group animals for a reason. If man is wolf to man, he is so in every sense, not just the negative one. We would not be where we are today had our ancestors been socially aloof.

What we need is a complete overhaul of assumptions about human nature. Too many economists and politicians model human society on the perpetual struggle they believe exists in nature, but which is a mere projection. Like magicians, they first throw their ideological prejudices into the hat of nature, then pull them out by their very ears to show how much nature agrees with them. It's a trick we have fallen for for too long. Obviously, competition is part of the picture, but humans can't live by competition alone.

The Over-kissed Child

The German philosopher Immanuel Kant saw as little value in human kindness as former U.S. vice president Dick Cheney did in energy conservation. Cheney mocked conservation as "a sign of personal virtue" that, sadly, wouldn't do the planet any good. Kant praised compassion as "beautiful" yet considered it irrelevant to a virtuous life. Who needs tender feelings if duty is all that matters?

We live in an age that celebrates the cerebral and looks down upon emotions as mushy and messy. Worse, emotions are hard to control, and isn't self-control what makes us human? Like hermits resisting life's temptations, modern philosophers try to keep human passions at arm's length and focus on logic and reason instead. But just as no hermit can avoid dreaming of pretty maidens and good meals, no philosopher can get around the basic needs, desires, and obsessions of a species that, unfortunately for them, actually is made of flesh and blood. The notion of "pure reason" is pure fiction.

If morality is derived from abstract principles, why do judgments often come instantaneously? We hardly need to think about them. In fact, psychologist Jonathan Haidt believes we arrive at them intuitively. He presented human subjects with stories of odd behavior (such as a one-night stand between a brother and sister), which the subjects immediately disapproved of. He then challenged every single reason they could come up with for their rejection of incest until his subjects ran out of reasons. They might say that incest leads to abnormal offspring, but in Haidt's story the siblings used effective contraception, which took care of this argument. Most of his subjects quickly reached the stage of "moral dumbfounding": They stubbornly insisted the behavior was wrong without being able to say why.

Clearly, we often make snap moral decisions that come from the "gut." Our emotions decide, after which our reasoning power tries to catch up as spin doctor, concocting plausible justifications. With this dent in the primacy of human logic, pre-Kantian approaches to morality are making a comeback. They anchor morality in the so-called sentiments, a view that fits well with evolutionary theory, modern neuroscience, and the behavior of our primate relatives. This is not to say that monkeys and apes are moral beings, but I do agree with Darwin, who, in *The Descent of Man,* saw human morality as derived from animal sociality:

> Any animal whatever, endowed with well-marked social instincts . . .
> would inevitably acquire a moral sense or conscience, as soon as its
> intellectual powers had become as well developed, or nearly as well
> developed, as in man.

What are these social instincts? What is it that makes us care about the behavior of others, or about others, period? Moral judgment obviously goes further than this, but an interest in others is fundamental. Where would human morality be without it? It's the bedrock upon which everything else is constructed.

Much occurs on a bodily level that we rarely think about. We listen to someone telling a sad story, and unconsciously we drop our shoulders, tilt our head sideways like the other, copy his or her frown, and so on. These bodily changes in turn create the same dejected state in us as we perceive in the other. Rather than our head getting into the other's head, it's our body that maps the other's. The same applies to happier emotions. I remember one morning walking out of a restaurant and wondering why I was whistling to myself. How did I get into such a good mood? The answer: I had been sitting near two men, obviously old friends, who hadn't seen each other in a long time. They had been slapping each other's backs, laughing, relating amusing stories. This must have lifted my spirit even though I didn't know these men and hadn't been privy to their conversation.

Mood transfer via facial expressions and body language is so powerful that people doing it on a daily basis literally start to look alike. This has been tested with portraits of longtime couples: One set of pictures was taken on their wedding day and another set twenty-five years later. Presented with separate portraits of these men and women, human subjects were asked to match them on similarity. For the set taken at an older age, they had no trouble deciding who was married to whom. But for the pictures taken at a younger age, subjects flunked the task. Married couples resemble each other, therefore, not because they pick partners who look like them, but because their features converge over the years. The similarity was strongest for couples who reported the greatest happiness. Daily sharing of emotions apparently leads one partner to "internalize" the other, and vice versa, to the point that anyone can see how much they belong together.

I can't resist throwing in here that dog owners and their pets also sometimes look alike. But this isn't the same. We can correctly pair photographs of people and their dogs only if the dogs are purebreds. It doesn't work with mutts. Purebreds, of course, are carefully selected by their owners, who pay high sums for them. An elegant lady may want to walk a wolfhound, whereas an assertive character may prefer a rottweiler. Since similarity doesn't increase with the number of years that owners have had their pets, the critical factor is the choice of breed. This is quite different from the emotional convergence between spouses.

Our bodies and minds are made for social life, and we become hopelessly depressed in its absence. This is why next to death, solitary confinement is our worst punishment. Bonding is so good for us that the most reliable way to extend one's life expectancy is to marry and stay married. The flip side is the risk we run after losing a partner. The death of a spouse often leads to despair and a reduced will to live that explains the car accidents, alcohol abuse, heart disease, and cancers that take the lives of those left behind. Mortality remains elevated for about half a year following a spouse's death. It is worse for younger than older people, and worse for men than women.

For animals, things are no different. I myself have lost two pets this way. The first was a jackdaw (a crowlike bird) that I had reared by hand. Johan was tame and friendly, but not attached to me. The love of his life was a female of his species, named Rafia. They were together for years, until Rafia one day escaped from the outdoor aviary (I suspect that a neighbor child had gotten curious and unlatched the door). Left behind, Johan spent days calling and scanning the sky. He died within weeks.

And then there was our Siamese cat, Sarah, who had been adopted as a kitten by our big tomcat, Diego, who would lick and clean her, let her knead his tummy as if she were nursing, and sleep with her. For about a decade they were best buddies, until Diego died of old age. Even though Sarah was younger and in perfect health, she stopped eating and died two months after Diego for no reason that the veterinarian could determine.

There exist of course thousands of such stories, including of animals that refuse to let go of loved ones. It is not unusual for primate mothers to carry their dead infants around until there's nothing left of them but skin and bones. A baboon female in Kenya who had recently lost her infant got extremely agitated when a week later she recognized the same bush on the savanna where she'd left its body. She climbed a high tree from which to scan while uttering plaintive calls normally used by baboons separated from their troop. Elephants, too, are known to return to the remains of dead companions to solemnly stand over their sun-bleached bones. They may take an hour to gently turn the bones over and over, smelling them. Sometimes they carry off bones, but other elephants have been seen returning them to the "grave" site.

Impressed by animal loyalty, humans have dedicated statues to it. In Edinburgh, Scotland, there's a little sculpture of "Greyfriars Bobby," a Skye terrier who refused to leave the grave of his master, buried in 1858. For fourteen whole years, Bobby guarded the grave while being fed by his fans, until he died and was buried not far away. His headstone reads "Let his loyalty and devotion be a lesson to us all." A similar statue exists in Tokyo for an Akita dog named Hachiko, who every day used to come to Shibuya Station to greet his master returning from work. The dog became famous for continuing this habit after his master had died in 1925. For eleven years, Hachiko waited at the appropriate time at the station. Dog lovers still gather once a year at the exit, now named after Hachiko, to pay homage to his faithfulness.

Touching stories, one might say, but what do they have to do with human behavior? The point is that we are mammals, which are animals with obligatory maternal care. Obviously, bonding has incredible survival value for us, the most critical bond being the one between mother and offspring. This bond provides the evolutionary template for all other attachments, including those among adults. We shouldn't be surprised, therefore, if humans in love tend to regress to the parent-offspring stage, feeding each other tidbits as if they can't eat by themselves, and talking nonsense with the same high-pitched voices normally reserved for babies. I myself grew up with the Beatles' love song lyrics "I wanna hold your hand"—another regression.

One set of animal studies has, in fact, had a huge, concrete influence on how humans treat one another. A century ago, foundling homes and orphanages followed the advice of a school of psychology that, in my opinion, has wreaked more havoc than any other: *behaviorism*. Its name reflects the belief that behavior is all that science can see and know, and therefore all it should care about. The mind, if such a thing even exists, remains a black box. Emotions are largely irrelevant. This attitude led to a taboo on the inner life of animals: Animals were to be described as machines, and students of animal behavior were to develop a terminology devoid of human connotations. Ironically, this advice backfired with at least one term. *Bonding* was originally coined to avoid anthropomorphic labels for animals, such as *friends* or *buddies*. But the term has since become so popular for human relationships (as in "male bonding," or "bonding experience") that now we probably will have to drop it for animals.

That humans are controlled by the same law-of-effect as animals was convincingly demonstrated by the father of behaviorism, John Watson, who inculcated in a human baby a phobia for hairy objects. At first, "Little Albert" happily played with the white rabbit he had been given. But after Watson paired each appearance of the rabbit with the loud clanging of steel objects right behind poor Albert's head, fear was the inevitable outcome. From then on, Albert placed his hands over his eyes and whimpered each time he saw the rabbit (or the investigator).

Watson was so enamored by the power of conditioning that he became allergic to emotions. He was particularly skeptical of maternal love, which he considered a dangerous instrument. Fussing over their children, mothers were ruining them by instilling weaknesses, fears, and inferiorities. Society needed less warmth and more structure. Watson dreamed of a "baby farm" without parents so that infants could be raised according to scientific principles. For example, a child should be touched only if it has behaved incredibly well, and not with a hug or kiss, but rather with a little pat on the head. Physical rewards that are systematically meted out would do wonders, Watson felt, and were far superior to the mawkish rearing style of the average well-meaning mom.

Unfortunately, environments like the baby farm existed, and all we can say about them is that they were deadly! This became clear when psychologists studied orphans kept in little cribs separated by white sheets, deprived of visual stimulation and body contact. As recommended by scientists, the orphans had never been cooed at, held, or tickled. They looked like zombies, with immobile faces and wide-open, expressionless eyes. Had Watson been right, these children should have been thriving, but they in fact lacked all resistance to disease. At some orphanages, mortality approached 100 percent.

Watson's crusade against what he called the "over-kissed child," and the immense respect accorded him in 1920s public opinion, seem incomprehensible today, but explains why another psychologist, Harry Harlow, set out to prove the obvious, which is that maternal love matters . . . to monkeys. At a primate laboratory in Madison, Wisconsin, Harlow demonstrated that monkeys reared in isolation were mentally and socially disturbed. When put in a group they lacked the tendency, let alone the skill, to interact socially. As adults, they couldn't even copulate or nurse offspring. Whatever we now think of the ethics of Harlow's research, he proved beyond any doubt that deprivation of body contact is not something that suits mammals.

With time, this kind of research changed the tide and helped improve the fate of human orphans. Except, that is, in Romania, where President Nicolae Ceauşescu created an emotional gulag by raising thousands of newborns in institutions. The world got a reminder of the nightmare of deprivation-rearing when Ceauşescu's orphanages opened after the fall of the Iron Curtain. The orphans were incapable of laughing or crying, spent the day rocking and clutching themselves in a fetal position (strikingly similar to Harlow's monkeys), and didn't even know how to play. New toys were hurled against the wall.

Bonding is essential for our species, and it is what makes us happiest. And here I don't mean the sort of jumping-for-joy bliss that the French leader General Charles de Gaulle must have had in mind when he allegedly sneered that "happiness is for idiots." The pursuit of happiness written into the U.S. Declaration of Independence rather refers

to a state of satisfaction with the
life one is living. This is a measur-
able state, and studies show that
beyond a certain basic income,
material wealth carries remarkably little weight. The standard
of living has been rising steadily
for decades, but has it changed
our happiness quotient? Not at
all. Rather than money, success,

*Romania's orphans were raised according to
"scientific" principles that ignored emotional needs.*

or fame, time spent with friends and family is what does people the
most good.

We take the importance of social networks for granted to the
point that we sometimes overlook them. This happened to my team
of primate experts—even though we should have known better—
when we built a new climbing structure for our chimpanzees. We focused too much on the physical environment. For more than thirty
years, the apes had lived in the same outdoor enclosure, a large open
area equipped with metal jungle gyms. We decided to get large telephone poles and bolt them together into something more exciting.
During the construction, the chimps were locked up next to the
site. At first they were noisy and restless, but upon hearing the huge
machine that put in the poles, they turned silent for the rest of the
time: They could hear that this was serious business! The poles were
connected with ropes; we planted new grass, dug new drains, and
eight days later we were ready. The new structure was ten times taller
than the one we had before.

At least thirty workers of the field station came to watch the release. We even had a betting pool about which chimp would be the
first to touch wood, or climb to the top. These apes had not smelled or
touched wood for decades; some of them never had. As one might
imagine, the director of the primate center guessed that the highest-
ranking male and female would be the first, but we knew that male
chimps are no heroes. They are always busy improving their political

position, taking great risks in the process, but they literally get diarrhea of fear as soon as something new comes around the corner.

Standing in the tower overlooking the compound with all cameras running, we released the colony. The first thing that happened was unexpected. We were so enamored with our wonderful construction, which had taken so much sweat to cobble together in the summer heat, that we had forgotten that the apes had been locked up for days in separate cages, even separate buildings. The first minutes following the release were all about social connections. Some chimps literally jumped into each other's arms, embracing and kissing. Within a minute, the adult males were giving intimidation displays, with all their hair on end, lest anyone might have forgotten who was boss.

The chimps barely seemed to notice the new construction. Some of them walked right underneath it as if it were invisible. They seemed in denial! Until they noticed the bananas we had placed at strategic locations visible from the ground. The first ones to get into the structure were the older females, and, ironically, the very last chimp to touch wood was a female known as the group's bully.

As soon as the fruits had been collected and eaten, though, everyone left the structure. They clearly weren't ready for it. They gathered in the old metal jungle gym, which my students had tested out the day before, finding it most uncomfortable to sit on. But the chimps had known it all their lives, so they lazily lay around in it looking up at the Taj Mahal that we had erected next to it, as if it were an object to be studied rather than enjoyed. It was months before they spent significant amounts of time in the new climbing frame.

We had been blinded by our own proud achievement, only to be corrected by the apes, who reminded us of the basics. It made me think again of Immanuel Kant, because isn't this the problem with modern philosophy? Obsessed by what we consider new and important about ourselves—abstract thought, conscience, morality—we overlook the fundamentals. I'm not trying to belittle what is uniquely human, but if we ever want to understand how we got there, we will need to start thinking from the bottom up. Instead of fixating on the peaks of civilization, we need to pay attention to the foothills. The peaks glimmer in the sun, but it is in the foothills that we find most of what drives us, including those messy emotions that make us spoil our children.

Macho Origin Myths

It was a typical primate conflict over dinner in a fancy Italian restaurant: one human male challenging another—me—in front of his girlfriend. Knowing my writings, what better target than humanity's place in nature? "Name one area in which it's hard to tell humans apart from animals," he said, looking for a test case. Before I knew it, between two bites of delicious pasta, I replied, "The sex act."

Perhaps reminded of something unmentionable, I could see that this took him aback a little, but only momentarily. He launched into a great defense of passion as peculiarly human, stressing the recent origin of romantic love, the wonderful poems and serenades that come with it, while pooh-poohing my emphasis on the mechanics of *l'amore*, which are essentially the same for humans, hamsters, and guppies (male guppies are equipped with a penislike modified fin). He pulled a deeply disgusted face at these mundane anatomical details.

Alas for him, his girlfriend was a colleague of mine, who with great enthusiasm jumped in with more examples of animal sex, so we had the sort of dinner conversation that primatologists love but that embarrasses almost everyone else. A stunned silence fell at neighboring tables when the girlfriend exclaimed that "he had *such* an erection!" It was unclear if the reaction concerned what she had just said, or that she had indicated what she meant holding thumb and index finger only slightly apart. She was talking about a small South American monkey.

Our argument was never resolved, but by the time desserts arrived it fortunately had lost steam. Such discussions are a staple of my existence: I believe that we are animals, whereas others believe we are something else entirely. Human uniqueness may be hard to maintain when it comes to sex, but the situation changes if one considers airplanes, parliaments, or skyscrapers. Humans have a truly impressive capacity for culture and technology. Even though many animals do show some elements of culture, if you meet a chimp in the jungle with a camera, you can be pretty sure he didn't produce it himself.

But what about humans who have missed out on the cultural growth spurt that much of the world underwent over the last few thousand years? Hidden in far-flung corners, these people do possess all the hallmarks of our species, such as language, art, and fire. We can study how they survive without being distracted by the technological advances of today. Does their way of life fit widely held assumptions about humanity's "state of nature"—a concept with a rich history in the West? Given the way this concept figured in the French Revolution, the U.S. Constitution, and other historical steps toward modern democracy, it's no trivial matter to establish how humans may have lived in their original state.

A good example are the "Bushmen" of southwest Africa, who used to live in such simplicity that their lifestyle was lampooned in the 1980 movie *The Gods Must Be Crazy*. As a teenager, anthropologist Elizabeth Marshall Thomas went with her parents, also anthropologists, to the Kalahari Desert to live among them. Bushmen, also known as the San, are a small, lithe people who have carved out a very modest niche in a grassy, open ecosystem that for half of the year is so low on water that the few reliable waterholes seriously restrict human movement. They have lived this way for thousands and thousands of years, which is why Marshall Thomas titled her book on them *The Old Way*.

A Bushman mother offers a child a drink from an ostrich eggshell filled with water.

The old way includes minimal clothing made out of antelope hides, a modest grass shelter, a sharpened digging stick, and an ostrich eggshell to transport water on day trips. Shelters are built and rebuilt all the time by putting a few sticks into the ground, intertwining the top, and covering the frame with grass. It reminded Marshall Thomas of the way apes build one-night nests in the trees by quickly weaving a few branches together into a platform before they go to sleep. This way, they stay off the ground, where danger lurks.

When Bushmen travel, they walk in single file, with a man in the lead who watches out for fresh predator tracks, snakes, and other dangers. Women and children occupy safer positions. This, too, is reminiscent of chimpanzees, who at dangerous moments—such as when they cross a human dirt road—have adult males in the lead and rear, with females and juveniles in between. Sometimes the alpha male stands guard at the road until everyone has crossed it.

Our ancestors may have been higher on the food chain than most primates, but they definitely were not at the apex. They had to watch their backs. This brings me to the first false myth about our state of nature, which is that our ancestors ruled the savanna. How could this be true for bipedal apes that stood only four feet tall? They must have lived in terror of the bear-sized hyenas of those days, and the saber-toothed cats that were twice the size of our lions. As a result, they had to content themselves with second-rate hunting time. Darkness is the best cover, but like the Bushmen today, early human hunters likely opted for the heat of the day, when their prey could see them coming from miles away, because they had to leave the night to the "professional" hunters.

Lions are the supreme rulers of the savanna, as reflected in our "lion king" stories and the Bushmen's high regard for lions. Significantly, Bushmen never use their deadly poison arrows on these animals, knowing that this may start a battle they can't win. The lions leave them alone most of the time, but when for some reason the lions in some places become man-eaters, people have had no choice but to leave. Danger is so much on the Bushmen's mind that at night, while the others sleep, they keep their fire going, which means getting up to stoke it. If the glow-in-the-dark eyes of nightly predators are spotted, appropriate action will be taken, such as picking up a burning branch from the fire and waving it over one's head (making one look larger-than-life) while urging the predator in a calm but steady voice to go find something better to do. Bushmen do have courage, but pleading with predators hardly fits the idea of humans as the dominant species.

The old way must have been quite successful, though, for even in the modern world we still show the same tendency to come together for safety. At times of danger, we forget what divides us. This was visible, for example, after the 9/11 attack on the World Trade Center in New York, an unbelievably traumatic experience for those who lived through it. Nine months afterward, when asked how they saw relations between the races, New Yorkers of all races called those relations mostly good, whereas in foregoing years, they had called them mostly bad. The postattack feeling of "we're in this together" had fostered unity in the city.

These reflexes go back to the deepest, most ancient layers of our brain, layers that we share with many animals, not just mammals. Look at how fish, such as herring, swim in schools that tighten instantly when a shark or porpoise approaches. Or how schools turn abruptly in one silvery flash, making it impossible for the predator to target any single fish. Schooling fish keep very precise individual distances, seek out companions of the same size, and perfectly match their speed and direction, often in a fraction of a second. Thousands of individuals thus act almost like a single organism. Or look at how birds, such as starlings, swarm in dense flocks that in an instant evade an approaching hawk. Biologists speak of "selfish herds," in which each individual hides among a mass of others for its own security. The presence of other prey dilutes the risk for each one among them, not unlike the old joke about two men being chased by a bear: There's no need to run faster than the bear so long as you outrun your pal.

Even bitter rivals seek companionship at times of danger. Birds that in the breeding season fight one another to death over territory may end up in the same flock during migration. I know this tendency first-hand from my fish, each time I redo one of my large tropical aquariums. Many fish, such as cich-lids, are quite territorial, displaying with spread fins and chasing one an-other to keep their corner free of intruders. I clean my tanks out every cou-

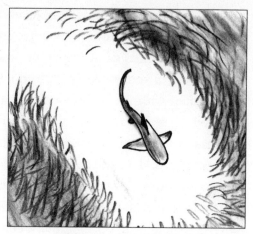

Fish band together in tight schools that confuse predators, such as these fish evading a shark.

ple of years, during which time I keep the fish in a barrel. After a few days they are released back into the tank, which by then looks quite different from before. I am always amused at how they suddenly seek out the company of their own kind. Like best buddies, the biggest fighters now swim side by side, exploring their new environment to-gether. Until, of course, they start to feel confident again, and claim a piece of real estate.

Security is the first and foremost reason for social life. This brings me to the second false origin myth: that human society is the volun-tary creation of autonomous men. The illusion here is that our ances-tors had no need for anybody else. They led uncommitted lives. Their only problem was that they were so competitive that the cost of strife became unbearable. Being intelligent animals, they decided to give up a few liberties in return for community life. This origin story, pro-posed by French philosopher Jean-Jacques Rousseau as the *social con-tract*, inspired America's founding fathers to create the "land of the free." It is a myth that remains immensely popular in political science departments and law schools, since it presents society as a negotiated compromise rather than something that came naturally to us.

Granted, it can be instructive to look at human relations *as if* they resulted from an agreement among equal parties. It helps us think about how we treat, or ought to treat, one another. It's good to realize, though, that this way of framing the issue is a leftover from pre-Darwinian days, based on a totally erroneous image of our species. As is true for many mammals, every human life cycle includes stages at which we either depend on others (when we are young, old, or sick) or others depend on us (when we care for the young, old, or sick). We very much rely on one another for survival. It is this reality that ought to be taken as a starting point for any discussion about human society, not the reveries of centuries past, which depicted our ancestors as being as free as birds and lacking any social obligations.

We descend from a long line of group-living primates with a high degree of interdependence. How the need for security shapes social life became clear when primatologists counted long-tailed macaques on different islands in the Indonesian archipelago. Some islands have cats (such as tigers and clouded leopards), whereas others don't. The same monkeys were found traveling in large groups on islands with cats, but in small groups on islands without. Predation thus forces individuals together. Generally, the more vulnerable a species is, the larger its aggregations. Ground-dwelling monkeys, like baboons, travel in larger groups than tree dwellers, which enjoy better escape opportunities. And chimpanzees, which because of their size have little to fear in the daytime, typically forage alone or in small groups.

Few animals lack a herd instinct. When former U.S. Senate majority leader Trent Lott titled his memoir *Herding Cats,* he was referring to the impossibility of reaching consensus. This may be frustrating when it comes to politicians, but for cats it's entirely logical. Domestic cats are solitary hunters, so don't need to pay much attention to one another. But all animals that either rely on one another for the hunt, such as members of the dog family, or are prey themselves, such as wildebeests, have a need to coordinate movements. They tend to follow leaders and conform to the majority. When our ancestors left the forest and entered an open, dangerous environment, they became prey and evolved a herd instinct that beats that of many animals. We excel at bodily synchrony and actually derive pleasure from it. Walking next to someone, for example, we automatically fall into the same stride. We coordinate chants and "waves" during sporting events, oscillate together during pop concerts, and take aerobics classes where we all jump up and down to the same beat. As an exercise, try to clap after a lecture when no one else is clapping, or try *not* to clap when everyone else is. We are group animals to a terrifying degree. Since political leaders are masters at crowd psychology, history is replete with people following them en masse into insane adventures. All that a leader has to do is create an outside threat, whip up fear, and voilà: The human herd instinct takes over.

Here we arrive at the third false origin myth, which is that our species has been waging war for as long as it has been around. In the 1960s, following the devastations of World War II, humans were routinely depicted as "killer apes"—as opposed to real apes, which were considered pacifists. Aggression was seen as the hallmark of humanity. While it's far from my intention to claim humans are angels of peace, we do need to draw a line between homicide and warfare. Warfare rests on a tight hierarchical structure of many parties, not all of which are driven by aggression. In fact, most are just following orders. Napoleon's soldiers didn't march into freezing Russia in an aggressive mood, nor did American soldiers fly to Iraq because they wanted to kill somebody. The decision to go to war is typically made by older men in the capital. When I look at a marching army, I don't necessarily see aggression in action. I see the herd instinct: thousands of men in lockstep, willing to obey superiors.

In recent history, we have seen so much war-related death that we imagine that it must always have been like this, and that warfare is written into our DNA. In the words of Winston Churchill, "The story of the human race is War. Except for brief and precarious interludes, there has never been peace in the world; and before history began, murderous strife was universal and unending." But is Churchill's warmongering state of nature any more plausible than Rousseau's noble savage? Although archeological signs of individual murder go back hundreds of thousands of years, we lack similar evidence for warfare (such as graveyards with weapons embedded in a large number of skeletons) from before the agricultural revolution. Even the walls of Jericho, considered one of the first pieces of evidence of warfare and famous for having come tumbling down in the Old Testament, may have served mainly as protection against mudflows.

Long before this, our ancestors lived on a thinly populated planet, with altogether only a couple of million people. Their density may have resembled that of the Bushmen, who live on ten square miles per capita. There are even suggestions that before this, about seventy thousand years ago, our lineage was at the edge of extinction, living in scattered small bands with a global population of just a couple of thousand. These are hardly the sort of conditions that promote continuous warfare. Furthermore, our ancestors probably had little worth fighting over, again like the Bushmen, for whom the only such exceptions are water and women. But Bushmen share water with thirsty visitors, and regularly marry off their children to neighboring groups. The latter practice ties groups together and means that the men in one group are often related to those in the other. In the long run, killing one's kin is not a successful trait.

Marshall Thomas witnessed no warfare among Bushmen and takes the absence of shields as evidence that they rarely fight with strangers. Shields, which are easily made out of strong hides, offer effective protection against arrows. Their nonexistence suggests that Bushmen are not too worried about intergroup hostilities. This is not to say that war is totally absent in preliterate societies: We know many tribes that engage in it occasionally, and some that do so regularly. My guess is that for our ancestors war was always a possibility, but that they followed the pattern of present-day hunter-gatherers, who do exactly the opposite of what Churchill surmised: They alternate long stretches of peace and harmony with brief interludes of violent confrontation.

Comparisons with apes hardly resolve this issue. Since it has been found that chimpanzees sometimes raid their neighbors and brutally take their enemies' lives, these apes have edged closer to the warrior

image that we have of ourselves. Like us, chimps wage violent battles over territory. Genetically speaking, however, our species is exactly equally close to another ape, the bonobo, which does nothing of
the kind. Bonobos can be unfriendly to their neighbors, but soon after
a confrontation has begun, females often rush to the other side to
have sex with both males and other females. Since it is hard to have sex
and wage war at the same time, the scene rapidly turns into a sort of
picnic. It ends with adults from different groups grooming each other
while their children play. Thus far, lethal aggression among bonobos
is unheard of.

The only certainty is that our species has a *potential* for warfare,
which under certain circumstances will rear its ugly head. Skirmishes
do sometimes get out of control and result in death, and young men
everywhere have a tendency to show off their physical prowess by
battling outsiders with little regard for the consequences. But at the
same time, our species is unique in that we maintain ties with kin long
after they have dispersed. As a result there exist entire networks between groups, which promote economic exchange and make warfare
counterproductive. Ties with outsiders provide survival insurance in
unpredictable environments, allowing the risk of food or water shortages to be spread across groups.

Polly Wiessner, an American anthropologist, studied "risk pool-
ing" among the Bushmen and offers the following description of the
delicate negotiations to obtain access to resources outside their territory. The reason these negotiations are done so carefully and indi-
rectly is that competition is never absent from human relations:

> In the 1970s, the average Bushman spent over three months a year
> away from home. Visitors and hosts engaged in a greeting ritual to
> show respect and seek permission to stay. The visiting party sat
> down under a shade tree at the periphery of the camp. After a few
> hours, the hosts would come to greet them. The visitors would tell
> about people and conditions at home in a rhythmic form of
> speech. The hosts would confirm each statement by repeating the

last words followed by "eh he." The host typically complained of food shortage, but the visitors could read how serious this was. If it was serious, they would say that they only had come for a few days. If the host did not stress shortages or problems, they knew they could stay longer. After the exchange, visitors were invited into camp where they often brought gifts, though they'd give them very subtly with great modesty so as not to arouse jealousy.

Because of interdependencies between groups with scarce resources, our ancestors probably never waged war on a grand scale until they settled down and began to accumulate wealth by means of agriculture. This made attacks on other groups more profitable. Instead of being the product of an aggressive drive, it seems that war is more about power and profit. This also implies, of course, that it's hardly inevitable.

So much for Western origin stories, which depict our forebears as ferocious, fearless, and free. Unbound by social commitments and merciless toward their enemies, they seem to have stepped straight out of your typical action movie. Present-day political thought keeps clinging to these macho myths, such as the belief that we can treat the planet any way we want, that humanity will be waging war forever, and that individual freedom takes precedence over community.

None of this is in keeping with the old way, which is one of reliance on one another, of connection, of suppressing both internal and external disputes, because the hold on subsistence is so tenuous that food and safety are the top priorities. The women gather fruits and roots, the men hunt, and together they raise small families that survive only because of their embeddedness in a larger social fabric. The community is there for them and they are there for the community. Bushmen devote much time and attention to the exchange of small gifts in networks that cover many miles and multiple generations. They work hard to reach decisions by consensus, and fear ostracism and isolation more than death itself. Tellingly, one woman confided, "It is bad to die, because when you die you are alone."

We can't return to this preindustrial way of life. We live in societies of a mind-boggling scale and complexity that demand quite a different organization than humans ever enjoyed in their state of nature. Yet, even though we live in cities and are surrounded by cars and computers, we remain essentially the same animals with the same psychological wants and needs.

The Other Darwinism

*I have received in a Manchester Newspaper a rather good squib,
showing that I have proved "might is right," & therefore that
Napoleon is right & every cheating Tradesman is also right.*

—CHARLES DARWIN, 1860

Long ago, American society embraced competition as its chief
organizing principle even though everywhere one looks—at work, in
the street, in people's homes—one finds the same appreciation of
family, companionship, collegiality, and civic responsibility as everywhere else in the world. This tension between economic freedom and
community values is fascinating to watch, which I do both as an out-
sider and an insider, being a European who has lived and worked in
the United States for more than twenty-five years. The pendulum
swings that occur at regular intervals between the main political parties of this nation show that the tension is alive and well, and that a
hands-down winner shouldn't be expected anytime soon.

This bipolar state of American society isn't hard to understand.
It's not that different from the situation in Europe, except that all

political ideologies on this side of the Atlantic seem shifted to the right. What makes American politics baffling is the way it draws upon biology and religion.

Evolutionary theory is remarkably popular among those on the conservative end of the spectrum, but not in the way biologists would like it to be. The theory figures like a secret mistress. Passionately embraced in its obscure persona of "Social Darwinism," it is rejected as soon as the daylight shines on real Darwinism. In a 2008 Republican presidential debate, no less than three candidates raised their hand in response to the question "Who doesn't believe in evolution?" No wonder that schools are hesitant to teach evolutionary theory, and that zoos and natural history museums avoid the e-word. Its hate-love relation with biology is the first great paradox of the American political landscape.

Social Darwinism is all about what Gordon Gekko called "the evolutionary spirit." It depicts life as a struggle in which those who make it shouldn't let themselves be dragged down by those who don't. This ideology was unleashed by British political philosopher Herbert Spencer, who in the nineteenth century translated the laws of nature into business language, coining the phrase "survival of the fittest" (often incorrectly attributed to Darwin). Spencer decried attempts to equalize society's playing field. It would be counterproductive, he felt, for the "fit" to feel any obligation toward the "unfit." In dense tomes that sold hundreds of thousands of copies, he said of the poor that "the whole effort of nature is to get rid of such, to clear the world of them, and make room for better."

The United States listened attentively. The business world ate it up. Calling competition a law of biology, Andrew Carnegie felt it improved the human race. John D. Rockefeller even married it with religion, concluding that the growth of a large business "is merely the working out of a law of nature and a law of God." This religious angle—still visible in the so-called Christian Right—forms the second great paradox. Whereas the book found in most American homes and every hotel room urges us on almost every page to show compassion,

Social Darwinists scoff at such feelings, which only keep nature from running its course. Poverty is dismissed as proof of laziness, and social justice as a weakness. Why not simply let the poor perish? I find it hard to see how Christians can embrace such a harsh ideology without a massive case of cognitive dissonance, but many seem to do so.

The third and final paradox is that the emphasis on economic freedom triggers both the best and worst in people. The worst is the aforementioned deficit in compassion, at least at the governmental level, but there is also a good, even excellent, side to the American character—otherwise I might have packed my bags long ago—which is a merit-based society. Silver spoons, fancy titles, family legacies, all of them are known and respected, but not nearly as much as personal initiative, creativity, and plain hard work. Americans admire success stories, and will never hold honest success against anyone. This is truly liberating for those who are up to the challenge.

Europeans are far more divided by rank and class and tend to prefer security over opportunity. Success is viewed with suspicion. It's not for nothing that the French language offers only negative labels for people who have made it by themselves, such as *nouveau riche* and *parvenu*. The result, in some nations, has been economic gridlock. When I see twenty-year-olds march in the streets of Paris to claim job protection or older people to preserve retirement at fifty-five, I feel myself all of a sudden siding with American conservatives, who detest entitlement. The state is not a teat from which one can squeeze milk any time of the day, yet that's how many Europeans seem to look at it.

And so my political philosophy sits somewhere in the middle of the Atlantic—not too comfortable a place. I appreciate the economic and creative vitality on this side but remain perplexed by the widespread hatred of taxes and government. Biology is very much part of this mix, as it is for every ideology that seeks justification. Social Darwinism sought to supply a scientific endorsement craved by a nation of immigrants who had quite naturally developed a strong sense of self-reliance and individualism.

The problem is that one can't derive the goals of society from the goals of nature. Trying to do so is known as the *naturalistic fallacy,* which is the impossibility of moving from how things are to how things ought to be. Thus, if animals were to kill one another on a large scale, this wouldn't mean we have to do so, too, any more than we would have an obligation to live in perfect harmony if animals were to do so. All that nature can offer is information and inspiration, not prescription.

Information is critical, though. If a zoo plans a new enclosure, it takes into account whether the species to be kept is social or solitary, a climber or a digger, nocturnal or diurnal, and so on. Why should we, in designing human society, act as if we're oblivious to the characteristics of our species? A view of human nature as "red in tooth and claw" obviously sets different boundaries to society than a view that includes cooperation and solidarity as part of our background. Darwin himself felt uncomfortable about the "right of the strongest" lessons that others, such as Spencer, tried to extract from his theory. This is why I'm tired, as a biologist, to hear evolutionary theory being trotted out as a prescription for society by those who aren't truly interested in the theory itself and all that it has to offer.

Enlightened Self-Interest

The idea of competition within the same species over the same resources appealed to Darwin and helped him formulate the concept of natural selection. He had read Thomas Malthus's influential 1798 essay on population growth, according to which populations that outgrow their food supply will automatically be cut back by hunger, disease, and mortality. Unfortunately, Spencer read the same essay and drew different conclusions. If strong varieties progress at the expense of inferior ones, this was not only how it *was,* Spencer felt, but how it *ought to be.* Competition was good, it was natural, and society as a whole benefited. He applied the naturalistic fallacy to a T.

Why did Spencer's ideas fall on such receptive ears? It seems to me that he was offering a way out of a moral dilemma that people were only just getting used to. In earlier times, the rich didn't need any justification to ignore the poor. With their blue blood, the nobility considered itself a different *breed*. They showed their contempt for manual labor by being wasp-waisted in the West or growing elongated fingernails in the East. Not that they felt absolutely no obligation toward those underneath them—hence the expression *noblesse oblige*—but they had no qualms living in opulence, feasting on meat, slurping fine wine, and driving around in gilded carriages, while the masses were close to starving.

All of this changed with the Industrial Revolution, which created a new upper crust, one that couldn't overlook the plight of others so easily. Many of them had belonged to the lower class only a few generations before: They evidently were of the same blood. So, shouldn't they share their wealth? They were reluctant to do so, though, and were thrilled to hear that there was nothing wrong with ignoring those who worked for them, that it was perfectly honorable to climb the ladder of success without looking back. This is how nature works, Spencer assured them, thus removing any pangs of conscience the rich might feel.

Add to this a peculiarity of American society, its debt to migration. To migrate across the globe takes a strong will and independence. I can relate to this, as I myself am an immigrant. It is a giant step to leave your friends and family behind, as well as your language, cuisine, music, climate, and so on. Migration is a gamble, and I did it on an impulse, as I'm sure many others did before.

Nowadays, it isn't such a big deal. With jet travel, telephone, and e-mail, it's easy to stay in touch. In the old days, however, people left on rickety ships, known as "coffin ships." Those who survived the storms and diseases arrived in an unknown land. They could be pretty sure they would never see their native country again or the people they'd been close to. Imagine saying goodbye to your parents while knowing that they will die without you, and that you might not even hear about it. Disproportionate numbers of novelty seekers and risk takers arrived in Canada, Australia, and America, making for a partially self-selected population. Self-selection works a bit like natural selection: The next generations inherit the same personality type—both genetically and culturally. Since the goal of every immigrant is to build a better life, the inevitable outcome is a culture revolving around individual achievement.

This was already clear to French political thinker and statesman Alexis de Tocqueville:

> In Europe we habitually regard a restless spirit, a moderate desire for wealth, and an extreme love of independence as great social dangers, but precisely these things assure a long and peaceful future in the American republics.

No wonder Spencer's message about success as its own justification was well received. More recently, a Russian American immigrant delivered the same message in a different package. Ayn Rand scoffed at the idea that success comes with moral obligations. She reached millions of enthusiastic readers with her message that egoism is no vice, but rather a virtue. She turned the whole issue upside down, devoting thousand-page novels to the notion that if we have any obligation at all, it is to ourselves. Former Federal Reserve chairman Alan Greenspan considers Rand a major influence on his life and work.

Insofar as such arguments are based on what is supposedly natural, however, they are fundamentally flawed. In Spencer's days, this was exposed by the unlikely character of a Russian prince, Petr Kropotkin. Though a bearded anarchist, Kropotkin was also a naturalist of great distinction. In his 1902 book, *Mutual Aid*, he argued that the struggle for existence is not so much one of each against all, but of masses of organisms against a hostile environment. Cooperation is common, such as when wild horses or musk oxen form a ring around their young to protect them against attacking wolves.

Kropotkin was inspired by a setting quite unlike the one that had inspired Darwin. Darwin visited tropical regions with abundant

A horned wall of adult musk oxen faces predators, such as wolves.

wildlife, whereas Kropotkin explored Siberia. The ideas of both men reflect the difference between a rich environment, resulting in the sort of population density and competition envisioned by Malthus, and an environment that is frozen and unfriendly most of the time. Having witnessed climatic calamities in which horses were scattered by the wind and herds of cattle perished under piles of snow, Kropotkin objected to the depiction of life as a "gladiator's show." Instead of animals duking it out, and the victors running off with the prize, he saw a communal principle at work. In subzero cold, you either huddle together or die.

Mutual aid has become a standard ingredient of modern evolutionary theories, albeit not exactly in the way Kropotkin formulated it. Like Darwin, he believed that cooperative groups of animals (or humans) would outperform less cooperative ones. In other words, the ability to function in a group and build a support network is a crucial survival skill. The importance of such skills for primates was confirmed by a recent baboon study on the Kenyan plains: Females with the best social ties were shown to have the most surviving infants. Grooming partners protect each other from outside aggression, send shrill warning calls to each other when they spot a predator, and provide soothing contact. All of this helps baboon mothers raise offspring.

I myself knew two inseparable female macaques named Ropey and Beatle. They were approximately the same age, and at first I thought they were sisters, because they did everything together, groomed each other, and gave friendly lip smacks to each other's babies. They also helped each other in fights, so much so that Beatle (who ranked below Ropey) would scream and look at her friend every time another monkey dared to threaten her. Everyone in the group knew that they would have to deal with both of them. According to our records, however, Ropey and Beatle were unrelated.

Theirs was just one of those trusting alliances that monkeys develop to get ahead. All primates have this tendency, and some even invest in the community as a whole. Instead of just focusing on their own position, they demonstrate group-oriented behavior. This is most evident in relation to social harmony. For example, Chinese golden monkeys live in harems of one male with several females. The male is much larger than the females and has a beautiful thick coat of orange hair. When his females quarrel, he positions himself between them until they stop, while calming tempers by turning from one to the other with a friendly facial expression or by combing the hair on each female's back with his fingers.

In chimpanzees, both males and females actively broker community relations. In a large zoo colony that I studied, females would occasionally disarm males who were gearing up for a display. Sitting with their hair erect, hooting and swaying from side to side, male chimps may take up to ten minutes before launching a charge. This gives a female time to go over to the angry male and pry open his hands to remove heavy branches and rocks. Remarkably, the males let them do so.

Females also bring males together if they seem incapable of reconciling after a fight. The males sit opposite each other, looking at each other only obliquely, and a female approaches one, then the other, until she has brought them together and then they groom each other. We have seen mediating females literally take a male by the arm to drag him toward his rival.

The males themselves also do a lot of conflict resolution. This is the task of top-ranking males, who will step in when disputes overheat. Most of the time a mere approach with an imposing posture

A male chimpanzee settles a female dispute over watermelons by standing between both parties with arms spread out until they stop screaming.

calms things down, but if necessary, the male will literally beat the contestants apart. Males who act as arbitrator usually don't take sides, and can be remarkably effective at keeping the peace. In all of these cases, primates show *community concern:* They try to ameliorate the state of affairs in the group as a whole.

A student of mine, Jessica Flack, investigated the effect of such behavior in a different primate: the pigtail macaque. These handsome monkeys, with short curly tails, have a reputation for being highly intelligent. In Southeast Asia, the muscular males are commonly employed as "farm hands," which may lead to startling encounters in city traffic. A man will drive by on a motorbike with a nonhuman passenger sitting upright like a real person on the backseat, legs dangling on both sides. He is on his way to work at a plantation. The monkeys have been trained to follow shouted commands from the ground while they are high up in a palm tree dislodging ripe coconuts, which their master collects under the tree and sells on the market.

Pigtail macaques normally live in groups in which high-ranking males, like chimp males, act as police: They step in to break up fights and maintain order. We worked with about eighty monkeys in a large open-air corral. For days on end, Jessica would sit in a tower in the heat of a Georgia summer, with water in one hand and a microphone in the other, to narrate thousands of social events. Like studies of so-called knockout mice, in which a gene is disabled to study its effects, ours was a knockout study in which we temporarily inactivated the police to see how the group would fare.

Every two weeks, we'd pick a day on which the top three males were removed in the morning and put back in in the evening. The males were kept in a building next to the corral. If a skirmish broke out in the group, monkeys sometimes ran to the door behind which the males sat to scream through the cracks, but obviously they'd have to work things out on their own that day. The effect of the knockouts was entirely negative: They produced increased fighting, more intense aggression, less reconciliation after fights, and a drop in grooming and play. On all measures, monkey society was falling apart.

A few individuals can make a huge difference: Social life benefits enormously from policing males. Note that the argument here is not that they sacrifice themselves for the group. All group-oriented behavior—mediation, disarmament, policing—serves the individuals who perform it. Females have an interest in reducing tensions among males, for whom it is not unusual to take their problems out on females and young. And males who are effective at keeping the peace often become enormously popular and respected in the group. But even so, group-oriented behavior improves the quality of the social environment not just for the individuals who show it, but for everyone else as well.

We often take collectivity for granted, but all group-living organisms are sensitive to it. They're all in the same boat. If this applies to other primates, how much more so to our own species with its more intricate societies? Most of us recognize the need to uphold certain services and institutions and are prepared to work toward this goal. Social Darwinists may disagree, but from a truly Darwinian perspective it is entirely logical to expect a "social motive" in group-living animals, one that makes them strive for a well-functioning whole.

By itself, this motive doesn't suffice. Perhaps bees or ants—that live in communities in which everyone is closely related and serves the same queen—are willing to work their hearts out for the common good, but humans are not. No matter how much brainwashing we engage in and patriotic songs we sing, we will always think of ourselves before we think of society. If any good has come out of the communist "experiment," it is this clarification of the limits of solidarity.

Purely selfish motives, on the other hand, don't suffice, either. There's such a thing as "enlightened" self-interest, which makes us work toward the kind of society that serves our own best interests. Both rich and poor rely on the same sewer system, highways, and law enforcement. All of us need national defense, education, and health care. A society operates like a contract: Those who gain from it are expected to contribute, and conversely, those who contribute feel entitled to get something out of it. We enter this contract automatically while growing up in a society, and react with outrage if it's violated.

At a 2007 political rally, Steve Skvara, a steelworker from Indiana, almost burst into tears describing his predicament:

> After 34 years with LTV Steel, I was forced to retire because of a disability. Two years later, LTV filed bankruptcy. I lost a third of my pension, and my family lost their health care. Every day of my life, I sit at the kitchen table across from the woman who devoted 36 years of her life to my family, and I can't afford to pay for her health care.

In the same way that Skvara felt an obligation to his wife, society ought to feel an obligation toward him after a lifetime of hard work. This is a *moral* issue, which is why Skvara received a standing ovation when he challenged the political candidates present, adding "What's wrong with America and what will you do to change it?"

In fact, American society is entering a period of correction, given the collapse of its financial system and the dimensions of its health-care crisis. Reliance on the profit principle has proven disastrous, so that the United States now ranks dead last in the industrialized world in terms of the quality of the health care that it provides. Western Europe, on the other hand, has enviable health care but it is, for other reasons and in other areas, moving in the opposite direction. When citizens are pampered by the state, they lose interest in economic advancement. They become passive players more interested in taking than in giving. Some nations have already turned back the clock on the welfare state, and others are expected to follow.

Every society needs to strike a balance between selfish and social motives to ensure that its economy serves society rather than the other way around. Economists often ignore this dynamic, thinking only in terms of money. Celebrated economist Milton Friedman claimed that "few trends could so very undermine the foundations of our free society as the acceptance by corporate officials of a social responsibility other than to make as much money for their stock-holders as possible." Friedman thus offered an ideology that puts people last.

Even if Friedman were right in theory about the connection be-tween money and freedom, in practice money corrupts. All too often it leads to exploitation, injustice, and rampant dishonesty. Given its colossal fraud, the Enron Corporation's sixty-four-page "Code of Ethics" now seems as fictional as the safety manual of the *Titanic*. In the past decade, every advanced nation has had major business scandals, and in every case executives have managed to shake the foundations of their society precisely by following Friedman's advice.

Enron and the Selfish Gene

Outside a hip restaurant I finally met my celebrity. My friends had promised that this place was frequented by Hollywood stars, and in-deed when darkness fell in the middle of dinner, and we spilled out onto the street, I found myself next to a cigarette-smoking movie idol whom I chatted with about this and that, and how our food must be getting cold. The encounter took place thanks to one of those rolling blackouts that struck California in 2000. Fifteen minutes later every-one was back at their table, back to normal, but of course what had just happened was extraordinary.

No, I don't mean meeting the star, but witnessing the wonders of unrestrained capitalism, all thanks to Enron, the Texas-based energy company that had developed innovative ways of tweaking the market and creating artificial power shortages so that prices would soar. Never mind that the blackouts posed serious risks for people on respirators or in elevators. Social responsibility just wasn't part of Enron's mind-set. They played by Friedman's rules but were inspired by an unexpected additional source that came straight out of the world of biology. The company's CEO, Jeff Skilling—now in prison—was a great fan of Richard Dawkins's *The Selfish Gene*, and deliberately tried to mimic nature by instigating cutthroat competition within his company.

Skilling set up a peer review committee known as "Rank & Yank." It ranked employees on a 1–5 scale of representing the best (1) or worst (5), and gave the boot to everyone ranked 5. Up to 20 percent of the employees were axed every year, but not without having been humiliated on a website featuring their portraits. They were first sent to "Siberia"—meaning that they had two weeks to find another position within the company. If they didn't, they were shown the door. The thinking behind Skilling's committee was that the human species has only two fundamental drives: greed and fear. This obviously turned into a self-fulfilling prophecy. People were perfectly willing to slit others' throats to survive within Enron's environment, resulting in a corporate atmosphere marked by appalling dishonesty within and ruthless exploitation outside the company. It eventually led to Enron's implosion in 2001.

The book of nature is like the Bible: Everyone reads into it what they want, from tolerance to intolerance, and from altruism to greed. It's good to realize, though, that if biologists never stop talking of competition, this doesn't mean they advocate it, and if they call genes selfish, this doesn't mean that genes actually are. Genes can't be any more "selfish" than a river can be "angry," or sun rays "loving." Genes are little chunks of DNA. At most, they are "self-promoting," because successful genes help their carriers spread more copies of themselves.

Like many before him, Skilling had fallen hook, line, and sinker for the selfish-gene metaphor, thinking that if our genes are selfish then we must be selfish, too. This is not necessarily what Dawkins meant, though, as became clear again during an actual debate that we had in a tower overlooking my chimpanzees.

As brief background, one needs to know that Dawkins and I had been critical of each other in print. He had said that I was taking poetic license with regard to animal kindness while I had chided him for coining a metaphor prone to be misunderstood. The usual academic bickering, perhaps, but serious enough that I feared some frost during our encounter at the Yerkes field station. Dawkins visited in connection with the production of a TV series, *The Genius of Charles Darwin*. The producers arrived ahead of him to set up a "spontaneous" encounter in which Dawkins would drive up to the door, step out of his van, walk toward me, shake my hand, and warmly greet me before we'd walk off together to see the primates. We did all of this as if it were the first time—even though we'd met before. To break the ice, I told him about the epic drought in Georgia, and how our governor had just led a prayer vigil on the steps of the state capitol to make sure we'd get some rain. This cheered up the staunch atheist, and we laughed at the marvelous coincidence that the vigil had been planned as soon as the weatherman had announced rain.

Our tower debate was frosty indeed, but only because it was one of those unusually chilly days in Georgia. With Dawkins unselfishly tossing fruits at the apes below, we quickly settled on common ground, which wasn't too hard given our shared academic background. I have no problems calling genes "selfish" so long as it's understood that this says *nothing* about the actual motives of humans or animals, and Dawkins agreed that all sorts of behavior, including acts of genuine kindness, may be produced by genes selected to benefit their carriers. In short, we agreed on a separation between what drives evolution and what drives actual behavior that is about as well recognized in biology as is the separation of church and state outside Georgia.

Overall, we had a splendid chat, as the Brits say, trying to flesh out this two-level approach. Before applying it here to kindness or altruism, let me start with a simpler example: color vision. Seeing colors is thought to have come about because our primate ancestors needed to tell ripe and unripe fruits apart. But once we could see color, the capacity became available for all sorts of other purposes. We use it to read maps, notice someone's blushing, or find shoes that match our blouse. This has little to do with fruits, although colors indicating

ripeness—red or yellow—still get us excited and are therefore prominent in traffic lights, advertisements, and works of art. On the other
hand, nature's default color—green—is considered calming, restful,
and boring.

The animal kingdom is full of traits that evolved for one reason
but are also used for others. The hoofs of ungulates are adapted to run
on hard surfaces, but also deliver a mean kick to pursuers. The hands
of primates evolved to grasp branches, but also help infants cling to
their mothers. The mouths of fish are made for feeding, but also serve
as "holding pens" for the fry of mouth-breeding cichlids. When it
comes to behavior, too, the original function doesn't always tell us
how and why a behavior will be used in daily life. Behavior enjoys
motivational autonomy.

A good example is sex. Even though our genital anatomy and sexual urges evolved for reproduction, most of us engage in sex without
paying attention to its long-term consequences. I've always thought
that the main impetus for sex must be pleasure, but in a recent poll by
American psychologists Cindy Meston and David Buss people offered
a bewildering array of reasons, from "I wanted to please my boyfriend"
and "I needed a raise" to "we had nothing to do" and "I was curious
how she'd be in bed." If humans usually engage in sex without giving
reproduction a thought—which is why we have the morning-after
pill—this holds even more for animals, which don't know the connection between sex and reproduction. They have sex because they
are attracted to one another, or because they have learned its pleasurable effects, but not because they want to reproduce. One can't want
something one doesn't know about. This is what I mean by motivational autonomy: The sex drive is hardly concerned with the reason
why sex exists in the first place.

Or consider the adoption of young that aren't one's own. If the
mother of a juvenile primate dies, other females often take care of it.
Even adult males may carry an unrelated orphan around, protecting
it and letting it remove food from their hands. Humans, too, adopt on
a large scale, often going through hellish bureaucratic procedures to

find a child to bestow care upon. The strangest cases, though, are cross-species adoptions, such as a canine bitch in Buenos Aires, Argentina, that became famous for having saved an abandoned baby boy by placing him alongside her own puppies in an act reminiscent of Romulus and Remus. This adoptive tendency is well-known at zoos, one of which had a Bengal tigress nurse piglets. The maternal instinct is remarkably generous.

Some biologists call such applications a "mistake," suggesting that behavior shouldn't be used for anything it wasn't intended for. Even if this sounds a bit like the Catholic Church telling us that sex isn't for fun, I can see their point. Instead of nursing those piggies, the biologically optimal thing for the tigress would have been to use them as protein snacks. But as soon as we move from biology to psychology, the perspective changes. Mammals have been endowed with powerful impulses to take care of vulnerable young, so that the tigress is only doing what comes naturally to her. Psychologically speaking, she isn't mistaken at all.

Similarly, if a human couple adopts a child from a faraway land, their care and worries are as genuine as those of biological parents. Or if people have sex because they "want to change the conversation" (an actual reason given in the above poll), their arousal and enjoyment are as real as that of any other couple. Evolved tendencies are part and parcel of our psychology, and we're free to use them any way we like.

Now, let's apply these insights to kindness. My main point is that even if a trait evolved for reason X, it may very well be used in daily life for reasons X, Y, and Z. Offering assistance to others evolved to serve self-interest, which it does if aimed at close relatives or group mates willing to return the favor. This is the way natural selection operates: It produces behavior that, on average and in the long run, benefits those showing it. But this doesn't mean that humans or animals only help one another for selfish reasons. The reasons relevant for evolution don't necessarily restrict the actor. The actor follows an existing tendency, sometimes doing so even if there's absolutely nothing to be gained: the man who jumps on the train tracks to protect a stranger, the dog who suffers massive injuries by leaping between a child and a rattlesnake, or the dolphins forming a protective ring around human swimmers in shark-infested water. It's hard to imagine that these actors are seeking future payoffs. Just as sex doesn't need to aim at reproduction, and parental care doesn't need to favor one's own offspring, assistance given to others doesn't require the actor to know if, when, and how he'll get better from it.

This is why the selfish-gene metaphor is so tricky. By injecting psychological terminology into a discussion of gene evolution, the two levels that biologists work so hard to keep apart are slammed together. Clouding of the distinction between genes and motivations has led to an exceptionally cynical view of human and animal behavior. Believe it or not, empathy is commonly presented as an illusion, something that not even humans truly possess. One of the most repeated quips in the sociobiological literature of the past three decades is "Scratch an 'altruist,' and watch a 'hypocrite' bleed." With great zeal and shock effect, authors depict us as complete Scrooges. In *The Moral Animal*, Robert Wright claims that "the pretense of selflessness is about as much part of human nature as is its frequent absence." The reigning incredulity concerning human kindness recalls a Monty Python sketch in which a banker is being asked for a small donation for the orphanage. Utterly mystified by the whole concept of a gift, the banker wonders "But what's my incentive?" He can't see why anyone would do anything for nothing.

Modern psychology and neuroscience fail to back these bleak views. We're preprogrammed to reach out. Empathy is an automated response over which we have limited control. We can suppress it, mentally block it, or fail to act on it, but except for a tiny percentage of humans—known as psychopaths—no one is emotionally immune to another's situation. The fundamental yet rarely asked question is: Why did natural selection design our brains so that we're in tune with our fellow human beings, feeling distress at their distress and pleasure at their pleasure? If exploitation of others were all that mattered, evolution should never have gotten into the empathy business.

At the same time, I should add that I have absolutely no illusions about the nasty side of our species, or that of any other primate, for that matter. I have witnessed more blood and gore among monkeys and apes than most. Too many times, I have watched vicious fights, seen males kill infants, or been left inspecting the wounds on a dead monkey, trying to determine if they were made by the sharp canine teeth of a male (slashes and punctures) or the smaller teeth of females (bruises and ripped skin). Aggression was my first topic of study, and I'm fully aware that there's no shortage of it in the primates.

It was only later that I became interested in conflict resolution and cooperation. The final push in this direction came from the death of my favorite chimp during the Machiavellian power struggles described in *Chimpanzee Politics*. Right before I emigrated, in 1980, two males at the Dutch zoo where I worked assaulted and castrated a third, named Luit, who later succumbed to his injuries. Similar incidents are now known from the field. I'm not referring here to the well-documented warfare over territory, which is directed against out-group members, but to the fact that wild chimps, too, occasionally kill within their own community.

Until this catastrophe, I had looked at conflict resolution as a mildly interesting phenomenon. I knew that chimpanzee contestants kiss and embrace each other after fights, but the shock of standing next to the veterinarian in a bloody operating room, handing him instruments for the hundreds of stitches he sewed, impressed upon me how critically important this behavior is. It helps apes maintain good relationships despite occasional conflict. Without these mechanisms, things get ugly. The tragic end of Luit opened my eyes to the value of peacemaking and played a major part in my decision to focus on what holds societies together.

The violent nature of chimps is sometimes used as an argument against their having any empathy at all. Since we associate empathy with kindness, a common question is "If chimps hunt and eat monkeys and kill their own kind, how can they possibly possess empathy?" What's most surprising is how rarely this question is being asked of our own species. If it were, we would of course be the first to disqualify as an empathic species. There exists in fact no obligatory connection between empathy and kindness, and no animal can afford treating everyone nicely all the time: Every animal faces competition over food, mates, and territory. A society based on empathy is no more free of conflict than a marriage based on love.

Like other primates, humans can be described either as highly co-operative animals that need to work hard to keep selfish and aggres-sive urges under control or as highly competitive animals that nevertheless have the ability to get along and engage in give-and-take. This is what makes socially positive tendencies so interesting: They play out against a backdrop of competition. I rate humans among the most aggressive of primates but also believe that we're masters at con-necting and that social ties constrain competition. In other words, we are by no means obligatorily aggressive. It's all a matter of balance: Pure, unconditional trust and cooperation are naïve and detrimental, whereas unconstrained greed can only lead to the sort of dog-eat-dog world that Skilling advocated at Enron until it collapsed under its own mean-spirited weight.

If biology is to inform government and society, the least we should do is get the full picture, drop the cardboard version that is Social Darwinism, and look at what evolution has actually put into place. What kind of animals are we? The traits produced by natural selection are rich and varied and include social tendencies far more conducive to optimism than generally assumed. In fact, I'd argue that biology constitutes our greatest hope. One can only shudder at the thought that the humaneness of our societies would depend on the whims of politics, culture, or religion.

Ideologies come and go, but human nature is here to stay.

3

Bodies Talking to Bodies

*When I'm watching an acrobat on a suspended wire,
I feel I'm inside of him.*

—THEODOR LIPPS, 1903

One morning, the principal's voice sounded over the intercom of my high school with the shocking announcement that a popular teacher of French had just died in front of his class. Everyone fell silent. While the headmaster went on to explain that the teacher had suffered a heart attack, I couldn't keep myself from having a laughing fit. To this day, I feel embarrassed.

What is it about laughter that makes it unstoppable even if triggered by inappropriate circumstances? Extreme bouts of laughter are worrisome: They involve loss of control, shedding of tears, gasping for air, leaning on others, even the wetting of pants while rolling on the floor! What a weird trick has been played on our linguistic species to express itself with stupid "ha ha ha!" sounds. Why don't we leave it at a cool "that was funny"?

These are ancient questions. Philosophers have been exasperated by the problem of why one of humanity's finest achievements, its sense of humor, is expressed with the sort of crude abandonment associated with animals. There can be no doubt that laughter is inborn. The expression is a human universal, one that we share with our closest relatives, the apes. A Dutch primatologist, Jan van Hooff, set out to learn under which circumstances apes utter their hoarse, panting laughs, and concluded that it has to do with a playful attitude. It's often a reaction to surprise or incongruity—such as when a tiny ape infant chases the group's top male, who runs away "scared," laughing all the while. This connection with surprise is still visible in children's games, such as peekaboo, or jokes marked by unexpected turns, which we save until the very end, appropriately calling them "punch lines."

Human laughter is a loud display with much teeth baring and exhalation (hence the gasping for air) that often signals mutual liking and well-being. When several people burst out laughing at the same moment, they broadcast solidarity and togetherness. But since such bonding is sometimes directed against outsiders, there is also a hostile element to laughter, as in ethnic jokes, which has led to the speculation that laughter originated from scorn and derision. I find this hard to believe, though, given that the very first chuckles occur between mother and child, where such feelings are the last things on their minds. This holds equally for apes, in which the first "playface" (as the laugh expression is known) occurs when one of the mother's huge fingers pokes and strokes the belly of her tiny infant.

The Correspondence Problem

What intrigues me most about laughter is how it *spreads*. It's almost impossible *not* to laugh when everybody else is. There have been laughing epidemics, in which no one could stop and some even died in a prolonged fit. There are laughing churches and laugh therapies based on the healing power of laughter. The must-have toy of

1996—Tickle Me Elmo—laughed hysterically after being squeezed three times in a row. All of this because we love to laugh and can't resist joining laughing around us. This is why comedy shows on television have laugh tracks and why theater audiences are sometimes sprinkled with "laugh plants": people paid to produce raucous laughing at any joke that comes along.

The infectiousness of laughter even works across species. Below my office window at the Yerkes Primate Center, I often hear my chimps laugh during rough-and-tumble games, and cannot suppress a chuckle myself. It's such a happy sound. Tickling and wrestling are the typical laugh triggers for apes, and probably the original ones for humans. The fact that tickling oneself is notoriously ineffective attests to its social significance. And when young apes put on their playface, their friends join in with the same expression as rapidly and easily as humans do with laughter.

Shared laughter is just one example of our primate sensitivity to others. Instead of being Robinson Crusoes sitting on separate islands, we're all interconnected, both bodily and emotionally. This may be an odd thing to say in the West, with its tradition of individual freedom and liberty, but Homo sapiens is remarkably easily swayed in one emotional direction or another by its fellows.

This is precisely where empathy and sympathy start—not in the higher regions of imagination, or the ability to consciously reconstruct how we would feel if we were in someone else's situation. It began much simpler, with the synchronization of bodies: running when others run, laughing when others laugh, crying when others cry, or yawning when others yawn. Most of us have reached the incredibly advanced stage at which we yawn even at the mere mention of yawning—as you may be doing right now!—but this is only after lots of face-to-face experience.

Yawn contagion, too, works across species. Virtually all animals show the peculiar "paroxystic respiratory cycle characterized by a standard cascade of movements over a five to ten second period" that defines the yawn. I once attended a lecture on involuntary *pandiculation* (the medical term for stretching and yawning) with slides of horses, lions, and monkeys—and soon the entire audience was pandiculating. Since it so easily triggers a chain reaction, the yawn reflex opens a window onto mood transmission, an essential part of empathy. This makes it all the more intriguing that chimpanzees yawn when they see others yawn.

This was first demonstrated at Kyoto University, where investigators showed apes in the laboratory the videotaped yawns of wild chimps. Soon the lab chimps were yawning like crazy. With our own chimps, we have gone one step further. Instead of showing them real chimps, we play three-dimensional animations of an apelike head going through a yawnlike motion. Devyn Carter, the technician who put these animations together, said he'd never yawned as much as during this particular job. Our apes also watch animations of a head merely opening and closing its mouth a couple of times, but they only yawn in response to the animated yawns. Their yawns look absolutely real, including maximal opening of the mouth, eye-closing, and head-rolling.

Yawn contagion reflects the power of unconscious synchrony, which is as deeply ingrained in us as in many other animals. Synchrony may be expressed in the copying of small body movements, such as a yawn, but also occurs on a larger scale, involving travel or movement. It is not hard to see its survival value. You're in a flock of birds and one bird suddenly takes off. You have no time to figure out what's going on: You take off at the same instant. Otherwise, you may be lunch.

Or your entire group becomes sleepy and settles down, so you too become sleepy. Mood contagion serves to coordinate activities, which is crucial for any traveling species (as most primates are). If my companions are feeding, I'd better do the same,

The animated yawns of an apelike head (similar to this one) induce real yawns in watching apes.

because once they move off, my chance to forage will be gone. The individual who doesn't stay in tune with what everyone else is doing will lose out like the traveler who doesn't go to the restroom when the bus has stopped.

The herd instinct produces weird phenomena. At one zoo, an entire baboon troop gathered on top of their rock, all staring in exactly the same direction. For an entire week, they forgot to eat, mate, or groom. They just kept staring at something in the distance that no one could identify. Local newspapers were carrying pictures of the monkey rock, speculating that perhaps the animals had been frightened by a UFO. But even though this explanation had the unique advantage of combining an account of primate behavior with proof of UFOs, the truth is that no one knew the cause except that the baboons clearly were all of the same mind.

The power of synchrony can be exploited for good purposes, as when horses were trapped on a piece of dry pasture in the middle of a flooded area in the Netherlands. Twenty horses had already drowned, and there were plenty of attempts to save the others. One of the more radical proposals was for the army to build a pontoon bridge, but before this was tried a far simpler solution came from the local horse riding club. Four brave women on horses mixed with the stranded herd, after which they splashed through a shallow area like pied pipers, drawing the rest with them. The horses walked most of the way, but had to swim a few stretches. In a triumph of applied animal knowledge, the riders reached terra firma followed by a single file of about one hundred horses.

Movement coordination both reflects and strengthens bonds. Horses that pull a cart together, for example, may become enormously attached. At first they jostle and push and pull against each other, each horse following its own rhythm. But after years of working together, the two horses end up acting like one, fearlessly pulling the cart at breakneck speed through water obstacles during cross-country marathons, complementing each other, and objecting to even the briefest separation as if they have become a single organism. The same principle operates among sled dogs. Perhaps the most extreme case was of a husky named Isobel, who after having turned blind still ran along perfectly with the rest based on her ability to smell, hear, and feel them. Occasionally, Isobel ran lead tandem with another husky.

In Dutch bicycle culture, it's common to have a passenger on the backseat. So as to follow the rider's movements, the person in the rear needs to hold on tightly—which is one reason that boys like to offer girls a ride. Bicycles turn not just by steering but also by leaning, so the passenger needs to lean the same way as the rider. A passenger who'd keep sitting up straight would literally be a pain in the behind. On motorcycles, this is even more critical. Their higher speed requires a deeper tilt in turns, and lack of coordination can be disastrous. The passenger is a true partner in the ride, expected to mirror the rider's every move.

Sometimes, a mother ape returns to a whimpering youngster who is unable to cross the gap between two trees. The mother first swings her own tree toward the one the youngster is trapped in, and then drapes her body between both trees as a bridge. This goes beyond mere movement coordination: It's problem-solving. The female is emotionally engaged (mother apes often whimper as soon as they hear their offspring do so), and adds an intelligent evaluation of the other's distress. Tree-bridging is a daily occurrence in traveling orangutans, in which mothers regularly anticipate their offspring's needs.

Even more complex are instances in which one individual takes charge of coordination between two others, as described by Jane Goodall with respect to three wild chimpanzees: a mother, Fifi, and her two sons. One son, Freud, had hurt his foot so badly that he was barely able to walk. Mother Fifi usually waited for him, but sometimes moved off before he was ready to limp after her. Her younger son, Frodo, proved more sensitive:

Three times when this happened Frodo stopped, looked from Freud to his mother and back, and began to whimper. He continued to cry until Fifi stopped once more. Then Frodo sat close to his big brother, grooming him and gazing at the injured foot, until Freud felt able to continue. Then the family moved on together.

This is not unlike my own personal experience. My mother has six sons, who all tower head and shoulders over her. Nevertheless, she has always been the leader of the pack. When she became older and frailer, however—which happened only in her late eighties—we had trouble adjusting. We'd step out of a car, for example, briefly help our mother out, but then walk briskly toward the restaurant or whatever place we were visiting, talking and laughing. We'd be called back by our wives, who'd gesture at our mother. She couldn't keep up and needed an arm to lean on. We had to adjust to this new reality.

Some of these examples are more complex than mere coordination: They involve assuming the perspective of someone else. Or, as in Goodall's and my family's account, alerting another to the situation of a third. The one thread that runs through all of these examples, however, is coordination. All animals that live together face this task, and synchrony is key. It is the oldest form of adjustment to others. Synchrony, in turn, builds upon the ability to map one's own body onto that of another, and make the other's movements one's own, which is exactly why someone else's laugh or yawn makes us laugh or yawn. Yawn contagion thus offers a hint at how we relate to others. Remarkably, children with autism are immune to the yawns of others, thus highlighting the social disconnect that defines their condition.

Body-mapping starts early in life. A human newborn will stick out its tongue in response to an adult doing so, and the same applies to monkeys and apes. In one research video, a tiny baby rhesus monkey intently stares at the face of an Italian researcher, Pier Francesco Ferrari, who slowly opens and closes his mouth several times. The longer the monkey watches the scientist, the more its own mouth begins to mimic his movements in a gesture that looks like the typical lip-smacking of its species. Lip-smacking signals friendly intentions and is as significant for monkeys as is the smile for humans.

I must say that I find neonatal imitation deeply puzzling. How does

A baby rhesus monkey stares at an experimenter and mimics his repeated mouth-opening.

a baby—whether human or not—mimic an adult? Scientists may bring up neural resonance or mirror neurons, but this hardly solves the mystery of how the brain (especially one as naïve as that of a neonate) correctly maps the body parts of another person onto its own body. This is known as the *correspondence problem:* How does the baby know that its own tongue, which it can't even see, is equivalent to the pink, fleshy, muscular organ that it sees slipping out from between an adult's lips? In fact, the word *know* is misleading, because obviously all of this happens unconsciously.

Body-mapping between different species is even more puzzling. In one study, dolphins mimicked people next to their pool without any training on specific behavior. A man would wave his arms, and the dolphins would spontaneously wave their pectoral fins. Or a man would lift up a leg, and the dolphins would raise their tails above the water. Think about bodily correspondence here, or in the case of a good friend of mine, whose dog started dragging her leg within days after he had broken his own. In both cases it was the right leg. The dog's limp lasted for weeks, but vanished miraculously once my friend's cast had come off.

As Plutarch said, "You live with a cripple, you will learn to limp."

The Art of Aping

Finding himself in front of the cameras next to his pal President George W. Bush, former British prime minister Tony Blair—known to walk normally at home—would suddenly metamorphose into a

distinctly un-English cowboy. He'd swagger with arms hanging loose and chest
puffed out. Bush, of course, strutted like
this all the time, and once explained how
back home, in Texas, this is known as
"walking."

Identification is the hook that draws
us in and makes us adopt the situation,
emotions, and behavior of those we're
close to. They become role models: We
empathize with them and emulate them.
Thus children often walk like the same-
sex parent, or mimic their tone of voice

*Children often emulate the
same-sex parent.*

when they pick up the phone. American playwright Arthur Miller
described how it works:

> Nothing was more enjoyable than mimicry. I was about the height
> of my father's back pocket, from which his handkerchief always
> hung out, and for years I pulled the corner of my handkerchief out
> exactly the same distance.

Imitation is also an anthropoid forte, as reflected in the verb "to
ape." Give a zoo ape a broom, and he'll move it across the floor the
way the caretaker does every day. Give her a rag and she'll soak it and
wring it out before applying it to a window. Hand him a key, and
you're in trouble! But even though all of this is common knowledge,
some scientists have been casting doubt on ape imitation. It just isn't
there, they say. Do these scientists have a point, or might they have
been testing their apes the wrong way?

In a typical experiment, an ape faces an unfamiliar white-coated
experimenter, who sits outside the cage to demonstrate a novel tool
that has no meaning in the ape's environment. After, say, five stan-
dardized demonstrations the tool is handed to the ape to see how
she'll use it. Never mind that apes dislike strangers, and that it's always

harder to relate to another species than one's own. Apes do poorly in these tests compared with children. But then again, the children aren't kept behind bars, and happily sit on their mothers' laps. They are being talked to, and most important, they're dealing with their own kind. They obviously feel perfectly at ease and have no trouble relating to the experimenter. Even though these studies seem to compare apples and oranges, they have fueled claims of a cognitive gap between apes and children.

Soon the inevitable happened: Imitation was elevated to a uniquely human skill. Never mind that such claims are always tricky, which is why they're being adjusted every couple of years, and that animals learn remarkably easily from companions. Examples range from birds or whales picking up songs from one another to the picnic wars between bears and people in the American wilderness. The bears invent new tricks all the time (they've learned, for example, that jumping up and down on top of a specific brand of car will pop open all of its doors), which then spread like wildfire through the population (resulting in warning signs at park entrances for the owners of these particular cars). Clearly, bears notice one another's successes. At the very least, human uniqueness claims should be downgraded to something more reasonable, such as that our imitation is more developed than that of other animals. But even then I'd be cautious, because our own research has fully restored faith in the aping skills of apes. By eliminating the human experimenter, we have gotten quite different results from the above studies. Given a chance to watch their own kind, apes copy every little detail they see.

Let me start with spontaneous imitation. Small infants in our chimpanzee colony sometimes get a finger stuck in the wire fence. Hooked the wrong way into the mesh, the finger cannot be extracted by force. Adults have learned not to pull at the infant, who always manages to free itself in the end. In the meantime, however, the entire colony has become agitated by the infant's screams: This is a rare but dramatic event analogous to a wild ape getting caught in a poacher's snare.

On several occasions, we have seen other apes mimic the victim's situation. The last time, for example, when I approached to assist I was greeted with threat barks from both the mother and the alpha male. As a result, I stayed back. One older juvenile came over to reconstruct the event for me. Looking me in the eyes, she inserted her finger into the mesh, slowly and deliberately hooking it around, and then pulling as if she too had gotten caught. Then two other juveniles did the same at a different location, pushing each other aside to get their fingers in the same tight spot they had found for this game. These juveniles themselves may long ago have experienced this situation for real, but now their charade was prompted by what had happened to the infant.

Our chimps obviously haven't read the scientific literature that says imitation is a way of reaching a goal or gaining rewards. They do so spontaneously, often without gains in mind. It's so much a part of their everyday life that I set up an ambitious research project together with a British colleague, Andy Whiten, who had been thinking along the same lines. In contrast to previous studies, we wanted to know how well apes learn from one another. From an evolutionary viewpoint, it doesn't really matter what they learn from us—all that matters is how they deal with their own kind.

To have one ape act as a model for another, however, is easier said than done. I can tell a co-worker to demonstrate a particular action and to repeat it ten times in a row, but try telling that to an ape! We faced an uphill battle, and our eventual success owes much to a rather "chimpy" young woman from Scotland, Vicky Horner. Mind you, "chimpy" isn't an insult for anyone who loves apes, and all I mean is that Vicky has the right body language (squatting down, no nervous movements, friendly disposition) and knows exactly which individuals act like divas, which ones demand respect, which just want to play and have fun, and which have eyes bigger than their bellies when there's food around. She deals with each personality on its own quirky basis, so that all of them feel at ease. If Vicky's rapport with apes was one weapon in our arsenal, the second was the rapport among the apes themselves. Most of our chimpanzees are either related or have grown up together, so they're more than willing to pay attention to one another. Like a close human family, they're one bickering and loving bunch, far more interested in one another than in us—the way apes ought to be.

Vicky employs the so-called two-action method. The apes get a puzzle box that they can access in two ways. For example, one either pokes a stick into it, and food rolls out, or one uses the stick to lift a lever, and food rolls out. Both methods work equally well. First, we teach the poking technique to one member of the group, usually a high-ranking female, and let her demonstrate it. The whole group gathers around to see how she gets her M&Ms. Then we hand the box over to her group mates, who obviously—if there is any truth to apes being imitators—should now favor the poking technique, too. This is indeed what they do. Next, we repeat our experiment at the same field station on a second group, which lives out of sight of the first. Here we teach another female the lifting technique, and lo and behold, her entire group develops a preference for lifting. We thus artificially create two separate cultures: "lifters" and "pokers."

The beauty of this outcome is that if chimps were to learn things on their own, each group should show a mix of solutions, not a bias for one or the other technique. Clearly, the example given by one of their group mates makes a huge difference. In fact, when we gave naïve chimps the same box, without any demonstrations, none of them was able to get any food out of it!

Next we tried a variation on the "telephone game" to see how information travels among multiple individuals. A new two-action box was built, one that could be opened by sliding a door to the side, or flipping the same door upward. We'd teach one individual to slide, after which another would watch the first, followed by a third watching the second, and so on. Even after six pairings, the last chimp still preferred sliding the door. Taking the same box to the other group, we produced an equally long chain that preferred the lifting solution.

Following the same procedures with human children in Scotland,

Andy obtained virtually identical results. I must admit to some jealousy, because with children such an experiment takes only a couple of days, whereas each time we set up a new experiment with our apes, we count on approximately a year to complete it. Our chimps live outdoors and participate on a volunteer basis. We call them by name, and just hope that they'll come in for testing (in fact, they know not only their own names, but also the others', which allows us to ask one chimp to fetch another). Adult males generally are too busy for our tests: Their power struggles and the need to keep an eye on one another's sexual adventures have priority. Females, on the other hand, have their reproductive cycles and offspring. If they come in alone, they may be very upset by the separation, which doesn't help our experiment, whereas if they do come with their youngest offspring, guess who will be playing with the box? That doesn't do us much good, either. If females are sexually attractive—sporting their balloonlike genital swellings—they may be willing to participate, but there will be three males who want to join incessantly banging on the door, thus killing all concentration. Or it could be that two chimps in a paired test have, unbeknownst to us, had a spat in the morning and now refuse to even look at each other. "It's always something," as we say, which explains why scientists have traditionally preferred setups in which apes interact with a human experimenter. This way, at least one party is under control.

Ape-to-ape testing is much harder but has huge payoffs. Allowed to imitate one another, apes entirely live up to their reputation. They're literally in one another's faces, leaning on one another, sometimes holding the model's hand while she's performing, or smelling her mouth when she's chewing the goodies she has won. None of this would be possible with a human experimenter, who is usually kept at a safe distance. Adult apes are potentially dangerous, which is why close personal contact with humans is prohibited. In order to learn from others, though, contact makes all the difference. Our chimps watch their model's every move, and often replicate the observed actions even before they've gained any rewards themselves. This means that they've learned purely from observation. This brings me back to the role of the body.

How does one chimp imitate another? Is it because he identifies with the other and absorbs its body movements? Or could it be that he doesn't need the other, and focuses on the box instead? Maybe all he needs to know is how the thing works. He may notice that the door slides to the side, or that something needs to be lifted up. The first kind of imitation involves reenactment of observed manipulations; the second merely requires technical know-how. Thanks to ingenious studies in which chimps were presented with a ghost box, we know which of these two explanations is correct. A ghost box derives its name from the fact that it magically opens and closes by itself so that no actor is needed. If technical know-how were all that mattered, such a box should suffice. But in fact, letting chimps watch a ghost box until they're bored to death—with its various parts moving and producing rewards hundreds of times—doesn't teach them anything.

In order to learn from others, apes need to see actual fellow apes: Imitation requires identification with a body of flesh and blood. We're beginning to realize how much human and animal cognition runs via the body. Instead of our brain being like a little computer that orders the body around, the body-brain relation is a two-way street. The body produces internal sensations and communicates with other bodies, out of which we construct social connections and an appreciation of the surrounding reality. Bodies insert themselves into everything we perceive or think. Did you know, for example, that physical condition colors perception? The same hill is assessed as steeper, just from looking at it, by a tired person than by a well-rested one. An outdoor target is judged as farther away than it really is by a person burdened with a heavy backpack than by one without it.

Or ask a pianist to pick out his own performance from among others he's listening to. Even if this is a new piece that the pianist has performed only once in silence (on an electronic piano without headphones on), he will be able to recognize his own play. While listening, he probably re-creates in his head the sort of bodily sensations that accompany an actual performance. He feels the closest match listening to himself, thus recognizing himself through his body as much as through his ears.

The field of "embodied" cognition is still very much in its infancy but has profound implications for how we look at human relations. We involuntarily enter the bodies of those around us so that their movements and emotions echo within us as if they're our own. This is what allows us, or other primates, to re-create what we have seen others do. Body-mapping is mostly hidden and unconscious but sometimes it "slips out," such as when parents make chewing mouth movements while spoon-feeding their baby. They can't help but act the way they feel their baby ought to. Similarly, parents watching a singing performance of their child often get completely into it, mouthing every word. I myself still remember as a boy standing on the sidelines of soccer games and involuntarily making kicking or jumping moves each time someone I was cheering for got the ball.

The same can be seen in animals, as illustrated in an old black-and-white photograph of Wolfgang Köhler's classic tool-use studies on chimpanzees. One ape, Grande, stands on top of wooden boxes that she has stacked up to reach bananas hung from the ceiling, while Sultan watches intently. Even though Sultan sits at a distance, he raises his arm in precise synchrony with Grande's grasping movement. Another example comes from a chimpanzee filmed while using a heavy rock as a hammer to crack nuts. The actor is being observed by a younger ape, who swings his own (empty) hand down in sync every time the first one

Sultan (sitting) making an empathetic grasping movement with his hand while watching Grande reach for bananas.

strikes the nut. Body-mapping provides a great shortcut to imitation.

Identification is even more striking at moments of high emotion. I once saw a chimpanzee birth in the middle of the day. This is unusual: Our chimps tend to give birth at night or at least when there are no humans around, such as during a lunch break. From my observation window I saw a crowd gather around Mai—quickly and silently, as if drawn by some secret signal. Standing half upright with her legs slightly apart, Mai cupped an open hand underneath of her, ready to catch the baby when it would pop out. An older female, Atlanta, stood next to her in similar posture and made exactly the same hand movement, but between *her own* legs, where it served no purpose. When, after about ten minutes, the baby emerged—a healthy son—the crowd stirred. One chimpanzee screamed, and some embraced, showing how much everyone had been caught up in the process. Atlanta likely identified with Mai because she'd had many babies of her own. As a close friend, she groomed the new mother almost continuously in the following weeks.

Similar empathy was described by Katy Payne, an American zoologist, for elephants:

> Once I saw an elephant mother do a subtle trunk-and-foot dance as she, without advancing, watched her son chase a fleeing wildebeest. I have danced like that myself while watching my children's performances—and one of my children, I can't resist telling you, is a circus acrobat.

Not only do we mimic those with whom we identify, but mimicry in turn strengthens the bond. Human mothers and children play games of clapping hands either against each other or together in the same rhythm. These are games of synchronization. And what do lovers do when they first meet? They stroll long distances side by side, eat together, laugh together, dance together. Being in sync has a bonding effect. Think about dancing. Partners complement each other's moves, anticipate them, or guide each other through their own movements. Dancing screams "We're in synchrony!" which is the way animals have been bonding for millions of years.

When a human experimenter imitates a young child's movements (such as banging a toy on a table or jumping up and down exactly like the child), he elicits more smiles and attention than if he shows the same infantile behavior independently of the child. In romantic situations, people feel better about dates who lean back when they lean back, cross their legs when they do, pick up their glass when they do, and so on. The attraction to mimicry even translates into money. The Dutch may be notoriously stingy, but tips at restaurants are twice as high for waitresses who repeat their clients' orders ("You asked for a salad without onions") rather than just exclaim "My favorite!" or "Coming up!" Humans love the sound of their own echo.

When I see synchrony and mimicry—whether it concerns yawning, laughing, dancing, or aping—I see social connection and bonding. I see an old herd instinct that has been taken up a notch. It goes beyond the tendency of a mass of individuals galloping in the same direction, crossing the river at the same time. The new level requires that one pay better attention to what others do and absorb how they do it. For example, I knew an old monkey matriarch with a curious drinking style. Instead of the typical slurping with her lips from the surface, she'd dip her entire underarm in the water, then lick the hair on her arm. Her children started doing the same, and then her grandchildren. The entire family was easy to recognize.

There is also the case of a male chimpanzee who had injured his fingers in a fight and hobbled around leaning on a bent wrist instead of his knuckles. Soon all of the young chimpanzees in the colony were walking the same way in single file behind the unlucky male. Like chameleons changing their color to match the environment, primates automatically copy their surroundings.

When I was a boy, my friends in the south of the Netherlands always ridiculed me when I came home from vacations in the north, where I played with boys from Amsterdam. They told me that I talked funny. Unconsciously, I'd return speaking a poor imitation of the harsh northern accent. The way our bodies—including voice, mood, posture, and so on—are influenced by surrounding bodies is one of the mysteries of human existence, but one that provides the glue that holds entire societies together. It's also one of the most underestimated phenomena, especially in disciplines that view humans as rational decision makers. Instead of each individual independently weighing the pros and cons of his or her own actions, we occupy nodes within a tight network that connects all of us in both body and mind.

This connectedness is no secret. We explicitly emphasize it in an art form that is literally universal. Just as there are no human cultures without language, there are none that lack music. Music engulfs us and affects our mood so that, if listened to by many people at once, the inevitable outcome is mood convergence. The entire audience gets uplifted, melancholic, reflective, and so on. Music seems designed for this purpose. I'm not necessarily thinking here of what music has become in Western concert halls with their stuffy, dressed-up audiences who aren't even tapping their feet lest they be considered undignified. But even these audiences experience mood convergence: Mozart's *Requiem* obviously affects a crowd differently than does a Strauss waltz. I'm thinking mostly of pop concerts at which thousands sing along with their idol while waving candles or cell phones through the air, or blues festivals, marching bands, gospel choirs, jazz funerals, even families singing "Happy Birthday," all of which permit a more visceral, bodily reaction to the music. At the end of a Christmas dinner in Atlanta, for instance, our whole table sang along melodramatically to *Elvis' Christmas Album.* The combination of great food, wine, friendship, and chant was intoxicating in more than one sense: We swung and laughed together, and ended up in the same spirit.

I once played piano in a band. It would be an understatement to say that we had little success, but I did learn that performing together requires role-taking, generosity, and being in tune—literally—to a degree found in few other endeavors. Our favorite song was "House of the Rising Sun" by The Animals, which we tried to invest with as much drama as we could. We felt the song's doom and gloom without knowing exactly what kind of house we were singing about, which I figured out only years later. What stuck with me, though, was the unifying effect of playing together.

Animal examples are not hard to come by, and here I don't just mean a howling pack of wolves, male chimpanzees hooting together to impress their neighbors, or the well-known dawn choruses of howler monkeys—said to be the loudest mammals on earth. I am referring to siamangs, which I heard for the first time in the jungles of Sumatra. Siamangs are large black gibbons who sing high up in the trees when the forest starts to heat up. It's a happy, melodious sound that touched me at a much deeper level than birdsong, probably because it is produced by a mammal. Siamang song is more full-bodied than that of any bird.

Their song usually starts with a few loud whoops, which gradually build into ever louder and more elaborate sequences amplified by their balloonlike throat sacs. Their sound carries for miles. At some point, the human listener correctly decides that a single animal can't be producing it. For many animals, it's the male's job to keep intruders out, but with siamangs—which live in small family groups—both sexes work toward this end. The female produces high-pitched barks, whereas the male often utters piercing screams that at short range will put every hair on your body on end. Their wild and raucous songs grow in perfect unison into what has been called "the most complicated opus sung by a land vertebrate other than man." At the same time that the duet communicates "Stay out!" to other members of their species, it also proclaims "We're one."

Like cart-pulling horses that work against each other before they work together, it takes time for siamangs to sing in harmony, and harmony may be critical to hold on to a partner or territory. Other siamangs can hear how close a pair is, and will move in if they discern discord. This is why German primatologist Thomas Geissmann noted: "Leaving a partner doesn't appear to be very attractive because the duets of fresh couples are noticeably poor." He found that couples that sang together a lot also spent more time together and synchronized their activities better.

One can literally tell a good siamang marriage by its song.

A Feeling Brain

When Katy Payne offered us the image of a human mother resonating with her acrobat child, she unwittingly used the same example

We identify with a high-wire artist to the point that we participate in every step he takes.

as the German psychologist responsible for the modern concept of empathy. We're in suspense watching a high-wire artist, said Theodor Lipps (1851–1914), because we vicariously enter his body and thus share his experience. We're on the rope with him. The German language elegantly captures this process in a single noun: *Einfühlung* (feeling into). Later, Lipps offered *empatheia* as its Greek equivalent, which means experiencing strong affection or passion. British and American psychologists embraced the latter term, which became "empathy."

I prefer the term *Einfühlung* since it conveys the movement of one individual projecting him- or herself into another. Lipps was the first to recognize the special channel we have to others. We can't feel anything that happens outside ourselves, but by unconsciously merging self and other, the other's experiences echo within us. We feel them as if they're our own. Such identification, argued Lipps, cannot be reduced to any other capacities, such as learning, association, or reasoning. Empathy offers direct access to "the foreign self."

How strange that we need to go back one century to learn about the nature of empathy in the writings of a long-forgotten psychologist. Lipps offered a bottom-up account, that is, one that starts from the basics, rather than the top-down explanations often favored by psychologists and philosophers. The latter tend to view empathy as a cognitive affair based on our estimation of how others might feel given how we would feel under similar circumstances. But can this explain the immediacy of our reactions? Imagine we're watching the fall of a circus acrobat and are capable only of empathy based on the recall of previous experiences. My guess is that we wouldn't react until the moment the acrobat lies in a pool of blood on the ground. But of course this is not what happens. The audience's reaction is absolutely instantaneous: Hundreds of spectators utter "ooh" and "aah" at the *very instant* that the acrobat's foot slips. Acrobats often perform such slips on purpose, without any intention of falling, precisely because they know how much their audience is with them every step of the way. I sometimes wonder where Cirque du Soleil would be without this instant connection.

Science is coming around to Lipps's position, but this wasn't the case yet when Swedish psychologist Ulf Dimberg began publishing on involuntary empathy in the early 1990s. He ran into stiff resistance from proponents of the more cognitive view. Dimberg demonstrated that we don't decide to be empathic—we simply are. Having pasted small electrodes onto his subjects' faces so as to register the tiniest muscle movements, he presented them with pictures of angry and happy faces on a computer screen. Humans frown in reaction to angry faces and pull up the corners of their mouths in reaction to happy ones. This by itself was not his most critical finding, however, because such mimicry could be deliberate. The revolutionary part was that he got the same reaction if the pictures flashed on the screen too briefly for conscious perception. Asked what they had seen after such a subliminal presentation, subjects knew nothing about happy or sad faces but had still mimicked them.

Expressions on a screen not only make our face muscles twitch, they also induce emotions. Those who had been exposed to happy faces reported feeling better than those who had been exposed to angry ones, even though neither group had any idea of what they had seen. This means that we're dealing with true empathy, albeit a rather primitive kind known as *emotional contagion*.

Lipps called empathy an "instinct," meaning that we're born with it. He didn't speculate about its evolution, but it's now believed that empathy goes back far in evolutionary time, much further than our species. It probably started with the birth of parental care. During 200 million years of mammalian evolution, females sensitive to their offspring outreproduced those who were cold and distant. When pups, cubs, calves, or babies are cold, hungry, or in danger, their mother needs to react instantaneously. There must have been incredible selection pressure on this sensitivity: Females who failed to respond never propagated their genes.

A good illustration is a female chimpanzee, named Krom, whom I knew at a zoo. Krom was fond of infants, and cared well enough for them so long as she could see them. Being deaf, however, she failed to respond to the soft yelps and whimpers of tiny infants in trouble, such as when they can't reach the nipple, are in danger of losing their grip on Mom's hair, or feel squeezed. I once saw Krom sit down on her infant and fail to get up when it burst out screaming. She reacted only upon noticing the worried reaction of *other* females. We ended up having another female adopt and raise this infant. Krom's case taught me how critically important it is for a mammalian female to be in tune with her offspring's every need.

Having descended from a long line of mothers who nursed, fed, cleaned, carried, comforted, and defended their young, we should not be surprised by gender differences in human empathy. They appear well before socialization: The first sign of emotional contagion—one baby crying when it hears another baby cry—is already more typical of baby girls than baby boys. Later on we see more gender differences. Two-year-old girls witnessing others in distress treat them with more concern than do boys of the same age. And in adulthood, women report stronger empathic reactions than men, which is one reason why women have been attributed a "tending instinct."

None of this denies male empathy. Indeed, gender differences usually follow a pattern of overlapping bell curves: Men and women differ on average, but quite a few men are more empathic than the average woman, and quite a few women are less empathic than the average man. With age, the empathy levels of men and women seem to converge. Some investigators even doubt that in adulthood there's much difference left.

Nevertheless, to seek the origin of empathy in parental care seems logical, which is why Paul MacLean drew attention to the calls of young mammals that are lost and want their mother back, known as "separation calls." The pioneering American neuroscientist, who in the 1950s first described the "limbic system," was interested in the origin of parental care. Young mammals call when lost or frightened, and their mother responds by picking them up. She's in a great hurry to take care of the problem, and if she's big and strong, you just don't want to get in her way (which is another human-versus-bear story). The evolution of attachment came with something the planet had never seen before: a feeling brain. The limbic system was added to the brain, allowing emotions, such as affection and pleasure. This paved the way for family life, friendships, and other caring relationships.

The central importance of social bonding is hard to deny. We have a tendency to describe the human condition in lofty terms, such as a quest for freedom or striving for a virtuous life, but the life sciences hold a more mundane view: It's all about security, social companionships, and a full belly. There is obvious tension between both views, which recalls that famous dinner conversation between a Russian literary critic and the writer Ivan Turgenev: "We haven't yet solved the problem of God," the critic yelled, "and you want to eat!"

Our nobler strivings come into play only once the baser ones have been fulfilled. If attachment and empathy are as fundamental as proposed, we had better pay close attention to them in any discussion of human nature. There is also no reason to expect these capacities only in humans. They should manifest themselves in any warm-blooded creature with hair, nipples, and sweat glands, which is part of what defines a mammal.

This obviously includes those pesky little rodents.

Commiserating Mice

I don't particularly enjoy telling this story, since it betrays prejudice, but I think anyone can understand why the Dutch, in the aftermath of World War II, were less than enamored with their neighbors to the east. As an undergraduate at the University of Nijmegen, I was taught by several German professors, who spoke Dutch with a thick accent. One of them was a grumpy old man, who was said to have been a concentration camp guard. Obviously, this couldn't be true, since he now would have been in jail, or worse, but this is what the rumor mill said.

To make matters worse, this professor manually killed the mice needed for our anatomy practicum. He didn't believe in death by ether, and would simply take a box with live mice and stand with his back turned to us. A few minutes later, a pile of dead mice with cracked necks lay on the counter.

In his defense, I must say that "cervical dislocation," as the practice is known, is probably quicker and more humane than other forms of euthanasia. But you can imagine that we found this professor a bit scary. And that's just us. How did the mice look at this procedure? The first mouse from the box didn't know what was coming, but what about the last one? Can rodents detect one another's pain? Do they feel one another's pain?

Before going any further, I must warn that reading up on the science of animal empathy can be a challenge for animal lovers. To see how animals react to the pain of others, investigators have often produced the pain themselves. I don't necessarily approve of these practices, and don't apply them myself, but it would be foolish to ignore the discoveries they've produced. The good news is that most of this research was carried out decades ago, and is unlikely to be repeated today.

In 1959, American psychologist Russell Church published a scientific paper under the provocative title "Emotional Reactions of Rats to the Pain of Others." Church trained rats to obtain food by pressing a lever, and found that if one rat noticed that lever-pressing shocked a neighboring rat, it would stop. This is remarkable. Why shouldn't the rat simply continue to get food and ignore its companion dancing in pain on an electric grid? Did these rats stop pressing because they were distracted, worried about their companion, or fearful for themselves?

The explanation offered by Church was typical of a time when conditioning was thought to underlie all behavior. He argued that a rat fears for its own well-being when it sees a companion in distress. But does an untrained rat have any reason to associate the squeals of others with pain to itself? The animals in the experiment had grown up in the laboratory with controlled temperature and light, ample food, and no predators. They had never encountered a situation like this before. It seems more likely that the sight, sound, or smell of another rat in pain arouses an innate emotional response. One rat's distress may simply distress another.

This study inspired a brief flurry of experiments on animal "empathy," "sympathy," and "altruism"—always put between quotation marks so as to avoid the wrath of behaviorists, who didn't believe in such concepts. This work was subsequently ignored, due partly to the taboo on animal emotions, and partly to the traditional emphasis on the nasty side of nature. As a result, animal studies are now seriously lagging behind what we know about human empathy. This may be changing, though, thanks to a new study by Canadian scientists, titled "Social Modulation of Pain as Evidence for Empathy in Mice." This time, the word *empathy* is free of quotation marks, reflecting the growing consensus that emotional linkage between individuals has the same biological basis in humans and other animals.

The news came too late for my old anatomy professor, but Jeffrey

Mogil, head of the pain lab at McGill University, almost felt as if his mice were talking to one another about their pain. He was puzzled time and again by the fact that when he was testing mice from the same cage, the order in which they were used seemed to affect their response. The last mouse showed more signs of pain than the first. One possibility is that the last mouse was sensitized by having seen others in pain. Mogil compared it to sitting in a dentist's lobby and seeing other patients coming out of the room after an obviously unpleasant experience. One can't help becoming primed for pain.

Pairs of mice were put through a pain test. Each mouse was placed in a transparent glass tube such that it could see the other. Either one or both mice were injected with diluted acetic acid, known to cause—in the words of the investigators—a mild stomachache. Mice respond to this treatment with stretching movements, suggesting discomfort. The basic finding was that a mouse would show more stretching with an injected partner, who was stretching, too, as opposed to a control partner. Since this applied only to mice that were cage mates, not to strangers, it couldn't be due to a simple negative association, because then the reaction should have been the same regardless of whether they knew each other. Further experiments explored which sense was involved by comparing anosmic mice (which lack olfaction), deaf mice, and mice that were prevented from seeing each other. Vision turned out to be critical: The reaction occurred only between mice that could see each other.

The mice showed pain contagion. That is, the sight of another in pain intensified their own pain response. Interestingly, in the presence of a stranger in pain, sensitivity went *down:* The mice became strikingly passive. This counterempathic reaction, however, was restricted to males, which are also potentially the most hostile to one another. Were they less than empathetic with their rivals?

This gender effect reminds me of how humans empathize with another's distress. Seeing the pain of a person we have just cooperated with activates pain-related areas in our own brains. This applies to both men and women. Yet in some studies the same procedure has been followed with partners who had been instructed to act unfairly in a game with the subject before the latter went into the brain scanner. Having been duped by someone, we show the opposite of empathy: At our seeing his pain, the brain's *pleasure* centers light up. We're getting a kick out of their misery! Such Schadenfreude occurs only in men, however, because women remain empathic. This may seem a typically human reaction, yet the underlying theme (male lack of empathy for potential rivals) resembles the mouse findings, and might well be a mammalian universal.

Finally, the investigators exposed pairs of mice to different sources of pain. One was the same acetic acid as before, but the second was a radiant heat source that might burn them if they came too close. Mice observing a cage mate in pain from the acid withdrew more quickly from the heat source, thus indicating heightened sensitivity to a completely different pain stimulus that required a different reaction. This precludes motor mimicry as an explanation: The mice seemed sensitized to pain in general. Any pain.

This study goes a long way toward reviving the tentative conclusions of the 1960s, showing that even with larger numbers of subjects and more rigorous methods, we get the same result: intensification of one's own experience based on the perceived reaction of others. This is close enough to "empathy" to call it that.

It's obviously not the imaginative kind of empathy that makes us truly understand how someone else feels, even someone we don't see, for example, when we read about the fate of a character in *War and Peace*. Yet it's good to keep in mind that imagination is not what drives empathy. Imagining another's situation can be a cold affair, not unlike the way we understand how an airplane flies. Empathy requires first of all emotional engagement. The mice show us how things may have gotten started. Seeing another's emotions arouses our own emotions, and from there we go on constructing a more advanced understanding of the other's situation.

Bodily connections come first—understanding follows.

Oscar the Cat

Oscar the Cat stares at us from a photograph in the prestigious *New England Journal of Medicine* along with an admiring description by a fellow expert. The author relates how Oscar makes his daily rounds at a geriatric clinic in Providence, Rhode Island, for patients with Alzheimer's, Parkinson's, and other illnesses. The two-year-old cat carefully sniffs and observes each patient, strolling from room to room. When he decides that someone is about to die, he curls up beside them, purring and gently nuzzling them. He leaves the room only after the patient has taken his or her last breath.

Oscar's predictions have been so dependable that the hospital staff counts on them. If he enters a room and leaves again, they know the patient's time isn't up yet. But as soon as Oscar starts one of his vigils, a nurse will pick up the phone to call family members, who then hurry to the hospital to be present while their loved one passes away. The cat has predicted the deaths of more than twenty-five patients with greater accuracy than any human expert. The tribute to the tomcat states: "No one dies on the third floor unless Oscar pays a visit and stays awhile."

How does Oscar do it? Is it the smell, skin color, or a certain pattern of breathing of dying patients? With so much variation in what ails them, it seems a bit unlikely that all patients end up showing the same telltale signs, but it's a possibility. Even more baffling is the question of what drives the cat. He has sometimes been the only one to be with an expiring patient, and the staff interprets this as him giving succor. But is this really what motivates our feline hospice?

I see two possible reasons for his behavior: It is either an attempt to comfort himself, if he's upset by what he senses is happening to a person, or an attempt to comfort the patient. But both possibilities remain perplexing. The first is so because it's unclear why Oscar would seek comfort with patients who have mostly become incapacitated: Wouldn't he be better off getting petted by some of the many people who'd love to do so? The second possibility is even harder to believe: Belonging to a species of solitary hunters, why would Oscar be so much more generous than any other cat I've ever known? I've had many in my lifetime, and whereas most cats do like to snuggle, I don't read much concern about our well-being in their behavior. To be perfectly cynical: I sometimes wonder why our cats love us so much more the colder it gets.

I'm exaggerating, of course. Cats do give affection and can show strong emotional connectedness. Otherwise, why would they always want to be in the same room we are in? The whole reason people fill their homes with furry carnivores and not with, say, iguanas or turtles—which are easier to keep—is that mammals offer us something no reptile ever will: emotional responsiveness. Dogs and cats have no trouble reading our moods and we have no trouble reading theirs. This is immensely important to us. We feel so much more at ease, so much more attached to animals with this capacity. Even if Oscar wasn't exactly acting out of concern, as I surmise, it would still be a mistake to dismiss his behavior as irrelevant to the issue of empathy.

Every evolved capacity is assumed to have advantages. If emotional contagion was indeed the first step on the road toward full-blown empathy, the question is, how does it promote survival and reproduction? The usual answer is that empathy produces helping behavior, but this hardly works for emotional contagion, which by itself doesn't do so. Take the typical reaction of a human toddler who hears another child cry. Her eyes fill with tears, upon which she runs to a parent to be picked up and comforted. In doing so, she in fact turns her back on the source of discomfort. Due to this lack of other-orientation, psychologists speak of "personal distress." It is a self-centered response that doesn't provide a good basis for altruism.

But that doesn't make emotional contagion useless. Let's say a wild rodent hears another squeal in fear and as a result becomes fearful itself. If this causes him to flee or go into hiding, he may avoid whatever fate befell the other. Or take a rodent mother, who is upset by her offsprings' ultrasonic distress peeps. She becomes restless herself, until she quiets her pups (and herself) by nursing them or moving them to a warmer spot. So without any deep interest in others' welfare, just by being emotionally aroused and reacting accordingly, animals may avoid danger or take care of their young. Things don't get any more adaptive than that.

The mother who "turns off" her pups' aversive noise by taking care of their problem is showing other-oriented behavior for self-centered reasons. I call this *self-protective altruism;* that is, helping another so as to shield oneself from aversive emotions. Such behavior does benefit others, yet lacks true other-orientation. Is this perhaps how concern for others evolved? Did it start with self-protective helping? Did this gradually evolve into helping geared toward the other's well-being? Libraries' worth of books try to draw a sharp line between selfishness and altruism, but what if we're facing an immense gray area? We can't exactly call empathy "selfish," because a perfectly selfish attitude would simply ignore someone else's emotions. Yet it doesn't seem appropriate either to call empathy "unselfish" if it is one's own emotional state that prompts action. The selfish/unselfish divide may be a red herring. Why try to extract the self from the other, or the other from the self, if the merging of the two is the secret behind our cooperative nature?

An intriguing example is how monkeys reacted in the same experiment discussed earlier for rats. In the 1960s, American psychiatrists reported that rhesus monkeys refused to pull a chain that delivered food to themselves if it shocked their companion. The monkeys went much further than the rats, which interrupted their behavior only briefly. One monkey stopped responding for five days and another for twelve days after witnessing the effect of its behavior on a companion. These monkeys were literally starving themselves to avoid inflicting pain on others.

Again, this was probably self-protective altruism: a desire to avoid unpleasant sights and sounds. It's just awful to watch others in pain, which is, of course, the whole point of empathy. Monkeys are extremely sensitive to one another's body language. This was shown in another experiment. One monkey watched a video screen that showed the face of another, who could hear a click sound that announced the arrival of electrical shocks for both. By deciphering the other's reaction to the sound, the first monkey could quickly press a lever that turned off the shocks. Even though the monkeys were sitting in separate rooms, they were highly successful at staying pain free. Apparently, the monkey with the lever had no trouble reading the face of the one who could hear the warning. The monkey was better at reading the other's expressions than the scientists who watched the same screen and concluded that "a monkey was a much more skilled interpreter of facial expression in another monkey than was man."

Isn't it horrible that such procedures are deemed necessary to prove the sensitivity of animals to one another? Can't research on animal empathy be conducted without arousing our own empathy? I'm not going to defend these procedures, but it's good to keep in mind how extremely little we know about animal empathy. Compared with the attention science has paid to negative emotions, such as fear and aggression, there has been a profound neglect of positive ones. It should be possible, however, to study empathy in a more benign way, as we do with humans. We could use mild stressors, for example, or reactions to spontaneous life events. After all, the daily life of primates is full of strains.

In my own research, I avoid causing pain or deprivation even though this leaves me with one obvious drawback: I never get to see what happens on the "inside" of my animals. Once, however, I saw a chance for an exception when radio transmitters became small enough to be implanted under the skin. This allowed measurement of a monkey's heart rate. It was being done with pets, so why not primates? In the old days, scientists needed to put monkeys in a restraining chair or outfit them with a heavy backpack to get heart data, but we were able to do so with a freely moving rhesus monkey. Live radio signals were picked up by an antenna mounted next to a young student, Stephanie Preston, who sat in a tower overlooking an outdoor corral with monkeys. We wanted to know how body contact affects the heart. I had just published my 1996 book, *Good Natured,* which had broached the controversial topic of animal empathy. How primates reduce stress was a big part of my argument, which is why we wanted to take a peek at their hearts.

In retrospect, I agree with one of my teachers, Robert Goy (the scientist who convinced me to move across the Atlantic), who warned me long ago: "Frans, stay away from the heart, because it's a mess." Obviously, he didn't mean the metaphorical heart of love and affection—he meant that it's almost impossible to make sense of heart rate. The heart reacts to everything: sex, aggression, fear, but also nonemotional activities such as jumping or running. Even if a monkey just sits up and scratches itself, its heart rate shoots up. How is one ever to figure out what's going on? If the heart slows down following a fight, for example, is this because the monkey is at peace, or is it simply because it stopped running and is now catching its breath?

At least we could tell that the monkey with the transmitter knew her relationship network intimately. If she was quietly sitting in the shade and another monkey strolled by, her heart rate would remain steady provided the other was a member of her family or a low-ranking monkey. Her heart would start racing, however, if the other was of high rank. We couldn't see much in her face or posture, but her heart revealed high anxiety. Rhesus monkeys live in the most hierarchical society I know, in which dominant individuals rarely hesitate to punish subordinates. They control them so completely that they sometimes even take food out of their mouths, literally—holding their heads still while reaching in. Our monkey's heart showed the silent terror that is life in rhesus society.

Stress begs alleviation, which rhesus monkeys achieve through grooming. The relaxing effect of this activity wasn't easy to prove, however, because for every time our monkey was being groomed we needed a perfect match, such as an almost identical situation in which she was *not* being groomed. The difference in heart rate could then be attributed to the grooming. We found indeed that grooming slows down the heart, which was the first such demonstration for any animal in a naturalistic setting. It confirmed the widely held assumption that grooming is an enjoyable, calming activity that serves not only to remove lice and ticks, but also to eliminate stress and foster social ties. Drops in heart rate have also been found in horses being petted by humans, and conversely, in humans petting their pets. In fact, animal companions are so effective against stress that they are increasingly recommended for heart patients.

I'll need to think of this the next time our cat, Sofie, wakes me up by tapping my face—ever so gently, but also ever so persistently—so she can slip under the covers.

In the winter, that is.

Empathy Needs a Face

During our heart rate study, Stephanie must have caught the empathy bug. After she'd gone on to study elsewhere, she decided to read more widely on the topic. The empathy literature is completely human-centered, never mentioning animals, as if a capacity so visceral and pervasive and showing up so early in life, could be anything other than biological. Empathy is still often presented as a voluntary process, requiring role-taking and higher cognition, even language. Stephanie and I wanted to go over the existing data from a different angle.

When I visited her years later in Berkeley, California, Stephanie dragged two large cardboard boxes from the corner of her office and put them on the table. I saw more articles on empathy than I had ever dreamed existed, neatly organized by topic, including historical papers, such as those by Theodor Lipps. Evidently, our review project had been growing larger and larger. The focus was on how empathy works, especially how the brain connects the outside world with the inside. The sight of another person's state awakens within us hidden memories of similar states that we've experienced. I don't mean conscious memories, but an automatic reactivation of neural circuits. Seeing someone in pain activates pain circuits to the point that we clench our jaws, close our eyes, and even yell "Aw!" if we see a child scrape its knee. Our behavior fits the other's situation, because it has become ours.

The discovery of *mirror neurons* boosts this whole argument at the cellular level. In 1992, an Italian team at the University of Parma first reported that monkeys possess special brain cells that fire not only when the monkey itself reaches for an object, but also when it sees another do so. In a typical demonstration, a computer screen shows the firing of a cell as recorded by electrodes in a monkey's brain. If the monkey takes a peanut from the experimenter's hand, the neuron gives a brief signal burst that (through an amplifier) sounds like a machine gun. When, a little later, the experimenter picks up a peanut while being watched by the monkey, the very same cell fires again. This time, however, it responds to *someone else's* action. What makes these neurons special is the lack of distinction between "monkey see" and "monkey do." They erase the line between self and other, and offer a first hint of how the brain helps an organism mirror the emotions and behavior of those around it. It's like a Pink Floyd song of long ago that draws attention to eye contact between people: "I am you and what I see is me." The discovery of mirror neurons has been hailed as being of the same monumental importance to psychology as the discovery of DNA has been for biology. That this key discovery took place in monkeys has obviously not helped claims of empathy as uniquely human.

The automaticity of empathy has become a point of debate, though. For the same reason that Dimberg ran into resistance showing unconscious facial mimicry, some scientists profoundly dislike any talk of automaticity, which they equate with "beyond control." We can't afford automatic reactions, they say. If we were to empathize with everybody in sight, we'd be in constant emotional turmoil. I'd be the last to disagree, but is this really what "automaticity" means? It refers to the speed and subconscious nature of a process, not the inability to override it. My breathing, for instance, is fully automated, yet I remain in charge. This very minute, I can decide to stop breathing until I see purple.

The ability to control and inhibit responses is not our only weapon against rampant empathy. We also regulate it at its very source by means of selective attention and identification. If you don't want to be aroused by an image, just don't look at it. And even though we identify easily with others, we don't do so automatically. For example, we have a hard time identifying with people whom we see as different or belonging to another group. We find it easier to identify with those like us—with the same cultural background, ethnic features, age, gender, job, and so on—and even more so with those close to us, such as spouses, children, and friends. Identification is such a basic precondition for empathy that even mice show pain contagion only with their cage mates.

If identification with others opens the door for empathy, the absence of identification closes that door. Since wild chimps occasionally kill one another, they must be capable of shutting the door completely. This takes place mostly when groups compete, which is of course also the situation in which humans run lowest on empathy. In one African reserve, a community of chimpanzees split into a northern and southern faction, eventually becoming two separate communities. These chimpanzees had played and groomed together, reconciled after squabbles, shared meat, and lived in harmony. But the factions began to fight over territory nonetheless. Shocked researchers watched as former friends literally drank one another's blood. Not even the oldest community members were exempt: An extremely frail-looking male was pummeled for twenty minutes, dragged about, and left for dead. This is why victims of chimpanzee warfare have been called "dechimpized," suggesting the same suppression of identification that marks dehumanization.

Empathy can also be nipped in the bud. Doctors and nurses in emergency rooms, for example, just cannot afford to be constantly in an empathic mode. They have to put a lid on it. There is a grisly side to this, such as the stories of Nazis who were quite sentimental about their own families, taking care of them as any normal father would, yet at the same time they had lamp shades made out of human skin and they exterminated masses of innocents. Or take Maximilien

Robespierre, the French revolutionary leader who rarely thought twice about sending "enemies of the Republic" to the guillotine—some of them former friends—yet loved to play with his dog, Brount, his sole companion on long walks. People who are perfectly attached and sensitive in one context may act like monsters in another.

But even if empathy is hardly inevitable, it is automatically aroused with those who have been "preapproved" based on similarity or closeness. With them, we can't help resonating. We often focus on the face, but obviously the entire body expresses emotions. As shown by Belgian neuroscientist Beatrice de Gelder, we react as rapidly to body postures as we do to facial expressions. We effortlessly read bodies, such as a fearful pose (ready to run, hands warding off danger) or an angry one (chest out, taking a step forward). When scientists played a trick on their subjects by pasting an angry face on the picture of a fearful body and a fearful face on an angry body, the incongruity slowed down reaction time. But the body posture won out when subjects were asked to judge the emotional state of the depicted person. Apparently, we trust postures more than facial expressions.

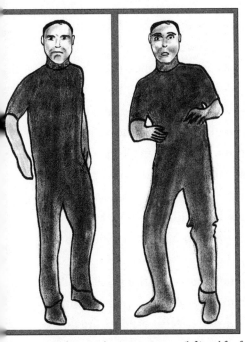

We show rapid reactions to angry (left) and fearful (right) body postures. In this drawing, the faces convey the same emotions as the bodies, but with the faces blacked out, we still show an emotional reaction purely based on posture.

How exactly the emotions of others affect our own is not entirely understood. One idea, which I'll call the "Body First Theory," holds that it starts with the body and that emotions follow. Someone else's body language affects our own body, which then creates an emotional echo that makes us feel accordingly. As Louis Armstrong sang, "When you're smiling, the whole world smiles with you." If copying another's smile makes us feel happy, the emotion of the smiler has been transmitted via our body. Strange as it may sound, this theory states that emotions arise from our bodies. For example, our mood can be improved by simply lifting up the corners of our mouth. If people are asked to bite down on a pencil lengthwise, taking care not to let the pencil touch their lips (thus forcing the mouth into a smile-like shape), they judge cartoons funnier than if they have been asked to frown. The primacy of the body is sometimes summarized in the phrase "I must be afraid, because I'm running."

This surely seems an odd way of putting things: Emotions are supposed to move us, not the other way around. Shouldn't it rather be "I run, because I'm afraid"? After all, "emotion" means to "stir" or "move." This is, in fact, the second idea, which I'll call the "Emotion First Theory." From seeing someone's body language or hearing their tone of voice, we deduce their emotional state, which then affects our own. In fact, we don't need to see their face to adopt the same facial expression, as has been demonstrated by letting humans watch pictures of fearful body postures with the faces blacked out. While this ruled out facial mimicry, the subjects' faces still registered fear. Emotional contagion thus relies on a direct channel between the other's and our own emotions.

There are times when matching the other's emotions is *not* a good idea. When we're facing a furious boss, for example, we'd get into deep trouble if we were to mimic his attitude. What we need is a quick grasp of his emotional state so as to respond with the appropriate submission, appeasement, or remorse. This applies almost equally to situations where the boss is right as where he is wrong. It's just a matter of social rank—a dynamic intuitively understood by every primate. The Emotion First Theory explains such encounters much better than the Body First Theory.

Despite the importance of body postures and movements, the face remains the emotion highway: It offers the quickest connection to the other. Our dependence on this highway may explain why people with immobile or paralyzed faces feel deeply alone, and tend to become depressive, sometimes to the point of suicide. Working with Parkinson's patients, a speech therapist noted that if in a group of, say, forty patients, five showed facial rigidity, all others would stay away from them. If they talked with them at all, it was to get simple "yes" or "no" answers. And if they wanted to know how they felt, they would rather speak with the companions of these patients. If empathy were a voluntary, conscious process of one mind trying to understand another, there would of course be no reason for this. People would simply need to put in a little more effort to hear the thoughts and feelings of these patients, who are perfectly capable of expressing themselves.

But empathy needs a face. With impoverished facial expression comes impoverished empathic understanding, and a bland interaction devoid of the bodily echoing that humans constantly engage in. As French philosopher Maurice Merleau-Ponty put it, "I live in the facial expression of the other, as I feel him living in mine." When we try to talk to a stone-faced person, we fall into an emotional black hole.

This is precisely the term used by a Frenchwoman who lost her face to a dog attack (her face had become nothing but a *grand trou*, she said, a "big hole"). In 2007, doctors gave her a new face, and her relief says it all: "I have returned to the planet of human beings. Those who have a face, a smile, facial expressions that permit them to communicate."

Someone Else's Shoes

Sympathy . . . cannot, in any sense, be regarded
as a selfish principle.

—ADAM SMITH, 1759

Empathy may be uniquely well suited for bridging the gap
between egoism and altruism, since it has the property of
transforming another person's misfortune into one's own
feeling of distress.

—MARTIN HOFFMAN, 1981

Walking into Moscow's State Darwin Museum, the very first display will surprise anyone familiar with the history of evolutionary thought. It's a life-size statue of Jean-Baptiste Lamarck, the French evolutionist whose ideas are often contrasted with those of Darwin himself.

Lamarck is depicted leaning back in an armchair with two teenage daughters standing by his side. The daughters look remarkably alike, and also resemble the bust, seen a little farther on, of Nadia Kohts, the Russian pioneer of animal studies, whose background I had come to investigate. The resemblance is no coincidence: Kohts posed for the sculptor of the statue. Photographs of her intelligent, dark-eyed face are featured in display after display in the museum: She was famous in Russia, and remains so today.

Even though we're used to primatological heroes of the female gender, the best-known among them have caught our attention by living close to dangerous creatures in the forest, defying the stereotype that only men would be brave enough to do so. Kohts was brave, too, but instead of lurking in the forest, the danger of her place and time resided in the Kremlin. Stalin, under the dark influence of his protégé, the amateur geneticist Trofim Lysenko, had many a brilliant biologist publicly recant their ideas, sent to the gulags, or quietly disappear. Names of the persecuted became unmentionable. Entire research institutes were closed down.

Thanks to its secular worldview, evolutionary theory was in favor with the Bolsheviks. Except, that is, for the idea of genetic change. Since this is a bit like accepting gravity without its pull, scientists had trouble with the tortuous way communism looked at evolution. Staying out of trouble became a major preoccupation for Kohts and her husband, Aleksandr, director of the museum. They hid their most sensitive documents and data among the stuffed animals in the basement, and made sure Lamarck received a prominent place in the museum. His theory, formulated before Darwin's, posited that acquired characteristics (such as the stretching of legs by wading birds or the lengthening of the giraffe's neck) can be passed on to the next generation. No genetic mutations are needed. The Lamarckian façade helped make the museum palatable to the powerful.

Kohts's isolation in Moscow had its advantages, though. She was oblivious to the doctrinal battles in the West, where scientists were busy closing the book on the animal mind. Acting as a surrogate mother for Yoni, a young chimpanzee, Kohts opened her heart and eyes to his every expression of sensitivity and intelligence. Rather than regarding him as a robot, devoid of thoughts and feelings, she saw him as a living being, not all that different from her own little son, Roody. She documented the development of her two charges in loving detail, being one of the first modern scientists to fully appreciate the emotional life of animals.

Kohts investigated Yoni's reactions to pictures of chimpanzees and other animals, to furs, and to his own reflection in a mirror. Even though Yoni was still too young to recognize what the mirror showed him, Kohts describes how, once he had gotten used to it, he would entertain himself by moving his tongue back and forth, and writhe and rotate it, closely studying its movements in the mirror. Kohts reports every aspect of Yoni's emotional development, from joy, jealousy, and guilt to sympathy and protection of loved ones. The following passage relates the extreme concern and compassion Yoni felt for Kohts:

> If I pretend to be crying, close my eyes and weep, Yoni immediately stops his play or any other activities, quickly runs over to me, all excited and shagged, from the most remote places in the house, such as the roof or the ceiling of his cage, from where I could not drive him down despite my persistent calls and entreaties. He hastily runs around me, as if looking for the offender; looking at my face, he tenderly takes my chin in his palm, lightly touches my face with his finger, as though trying to understand what is happening, and turns around, clenching his toes into firm fists.

What better evidence for the power of simian sympathy than the fact that an ape who'd refuse to descend from the roof of the house for food that was waved at him would do so instantly upon seeing his mistress in distress? Kohts describes how Yoni would look into her eyes when she pretended to cry: "the more sorrowful and disconsolate my crying, the warmer his sympathy." When she slapped her hands over her eyes, he tried to pull them away, extending his lips toward her face, looking attentively, slightly groaning and whimpering. She describes similar reactions from Roody, adding that her son went further than the ape in that he'd actually cry along with her. Roody cried even when he'd notice a bandage over the eye of his favorite uncle or when he'd see the maid grimace while swallowing bitter medicine.

The one limitation of Kohts's work was that she studied a single chimpanzee, and that he was so young. She never got to see the species' mature psychology and knew nothing about the way chimpanzees live in the wild. A psychologist who studied a single boy of a few years old would similarly be unable to generalize about the human species as a whole. On the other hand, because she was in contact with Yoni every day, and collected all possible information about him, Kohts was able to see a chimp up close in a way very few people have. She looked into the ape's heart and was impressed by what she saw.

Kohts included perceptive remarks about human behavior. For instance, when she sought a comparison for the temper tantrums that Yoni threw if he didn't get his way or was temporarily left alone, Kohts saw parallels gazing out the window of her study, which overlooked a morgue. Responding to the loss of a family member, especially in a case of accidental death, people utter heartbreaking cries while bending to the ground, almost under the wheels of the funeral carriage, making fitful, desperate gripping movements with their hands. She went on to comment on the human habit to gesticulate as a way of expressing and alleviating grief, comparing this to Yoni's hand gestures, which she found strikingly similar.

Having walked by Kohts's original writing desk in the museum, a photograph of her sitting next to her husband, another one in which the American expert of ape psychology Robert Yerkes talks with her via an interpreter, and a somber gallery of portraits honoring the many scientists executed under Lysenko and Stalin, I ran into a most unexpected display. Amid photos of Yoni laughing while being tickled and crying when frustrated, and an arrangement of his wooden toys and climbing ropes, stood Yoni himself. He has been preserved in a typical hooting posture—the way chimpanzees look when they are excited about something, such as food or company. The taxidermy is superb, as one would expect given that it was Aleksandr Kohts's specialty.

At first, I found it macabre to see the object of so much of Nadia Kohts's love and affection standing there as if still alive. But upon reflection, I concluded that preserving Yoni made sense for a couple as devoted to the traditional ways of natural history museums as the Kohtses were. After all, each had given the other a preserved animal as a wedding gift. For them, the best way to honor and commemorate Yoni must have been to make him part of their collection.

One of the greatest but least-known pioneers of primatology had left us her subject in an active pose, so that his obvious emotionality would catch our eye, as it did hers.

Sympathy

A monkey or rat reacting to another's pain by stopping the behavior that caused it may simply be "turning off" unpleasant signals. But self-protective altruism can't explain Yoni's reaction to his surrogate mother. First, because he hadn't caused her distress himself, and second, because he could easily have moved away when he saw her crying from the roof of the house. If self-protection had been his goal, he also should have left her hands where they were when she cried behind them. Clearly, Yoni wasn't just focusing on his own situation: He felt an urge to understand what the matter was with Kohts.

If Yoni were human, we'd speak of sympathy. Sympathy differs from empathy in that it is proactive. Empathy is the process by which we gather information about someone else. Sympathy, in contrast, reflects concern about the other and a desire to improve the other's situation. American psychologist Lauren Wispé offers the following definition:

> The definition of sympathy has two parts: first, a heightened awareness of the feelings of the other person, and, second, an urge to take whatever actions are necessary to alleviate the other person's plight.

Let me illustrate the distinction between sympathy and empathy by revealing something about myself: I have more empathy than sympathy. I'm not sure that this is a generalizable gender difference, but my wife seems to have equal amounts of both.

My profession depends on being in tune with animals. It would be terribly boring to watch them for hours without any identification, any intuition about what is going on, any ups and downs related to their ups and downs. Empathy is my bread and butter, and I have made many a discovery by closely following the lives of animals and trying to understand why they act the way they do. This requires that I get under their skin. I have no trouble doing so, love and respect animals, and do believe that this makes me a better student of their behavior.

But this is not sympathy. I have plenty of this as well, but it is less spontaneous, more subject to calculation, sometimes quite selfish. I am no Abraham Lincoln, who apparently interrupted a journey to pull a squealing pig from the mud. I don't necessarily stop for a lost dog or cat, whereas my wife, Catherine, picks up any stray she sees and works hard to locate its owner. If I know that one of my primates is gravely injured or ill—and under veterinary care—I am able to put it out of my head if I'm busy with something else. My mind is compartmentalized. Catherine worries without interruption about anyone who has fallen ill, whether human or animal, and will do anything within her power to take care of them. She's far more generous than I am. Perhaps I am more Kantian: thinking what's the right thing to do, weighing the pros and cons. Instead of flowing straight from my empathy, my sympathy takes a detour through a rational filter.

I recognize myself in a famous experiment, mischievously carried out on (male) seminary students. The students were ordered to walk to another building to lecture on the Good Samaritan, a religious outcast in a biblical parable who assists a man left for dead by the side of the road. On their way to the lecture hall, the students would pass a slumped person planted in an alleyway. The groaning "victim" would sit still with eyes closed and head down. Only 40 percent of the budding theologians asked what was wrong and offered assistance. Students who had been urged to make haste helped less than students with more time. Indeed, some students hurrying to explain the quintessential helping story of our civilization literally stepped over the stranger in need.

Thus, while empathy is easily aroused, sympathy is a separate process under quite different controls. It is anything but automatic. Nevertheless, it is common in both humans and other animals. When, in the 1970s, I first saw chimpanzees behave as solicitously as Yoni—albeit not to humans, but to one another—I labeled their behavior "consolation." You'd think that this was when my interest in empathy started, but instead of studying consolation in detail, I moved on. I was too fascinated by the way chimps make peace after fights with a kiss and embrace to pay attention to these other friendly contacts. It took me two decades to return to consolation, which happened when I realized how perfectly it fits the definitions of what psychologists call "sympathetic concern."

I have seen literally thousands of consolations—that's how common the behavior is. We have a massive computer database, compiled over many years, to tell us what happens after spontaneous fights among chimpanzees. Consolation is the most typical outcome. A victim of aggression, who not long ago had to run for her life, or scream to recruit support, now sits alone, pouting, licking an injury, or looking dejected. She perks up when a bystander comes over to give her a hug, groom her, or carefully inspect her injury. Consolations can be quite emotional, with both chimps literally screaming in each other's arms. Combing through the data to determine who shows consolation to whom, we found that it's mostly done by friends and relatives of the harmed party. Like Yoni, our chimps are sensitive to the plight of others, and go out of their way to alleviate suffering.

Ironically, this has been clear for a long time, but developments have conspired against it becoming widely known. First of all, until recently empathy was not taken seriously by science. Even with regard to our own species, it was considered an absurd, laughable topic classed with supernatural phenomena such as astrology and telepathy. A trailblazing child-empathy researcher once told me about the uphill battle to get her message across thirty years ago. Everything connected with empathy was seen as ill-defined, bleeding-heart kind of stuff, more suitable for women's magazines than hard-nosed science.

With regard to animals, the same resistance still exists. I had to think of this when seeing the picture of Yerkes chatting with Kohts, because those two were soul mates when it came to animal emotions. In one of his books, Yerkes complained how sympathy was the one topic he wasn't allowed to talk about despite his conviction that apes possessed it. He had often seen apes provide solace, even very young ones: "Impressive indeed is the thoughtfulness of the ordinarily carefree and irresponsible little chimpanzee for ill or injured companions." Yerkes rightly feared that he might be accused of idealizing animals if he told too many of these stories, especially about his favorite bonobo, Prince Chim. Of all the great apes, bonobos seem to have the highest level of empathy. In the 1920s, the species distinction between bonobos and chimpanzees was yet to be made, however, which is why Yerkes thought Prince Chim was just a special chimpanzee.

Of the many instances of bonobo sympathy known to me, perhaps the most remarkable one concerns a reaction to a bird. I've described this event before and normally would not repeat it here but there was an intriguing follow-up. The event concerns Kuni, who had found a stunned bird that had hit the glass wall of her zoo enclosure. Kuni took the bird up to the highest point of a tree to set it free. She spread its wings as if it were a little airplane, and sent it out into the air, thus showing a helping action geared to the needs of a bird. Obviously, such helping would not have worked for another bonobo, but for a bird it seemed perfectly appropriate. Kuni's reaction was probably based on what she knew about birds, seeing them fly by every day.

The parallel story that I recently heard concerned a bird, too. It happened at my old stomping ground, the Arnhem Zoo, where chimpanzees live on an island surrounded by a moat. The moat is full of life, such as fish, frogs, turtles, and ducks. One day, a couple of juvenile chimps had picked up a little duckling and were swinging it around, being far too rough with it, competing over who could play with it. When they tried to grab one of the other ducklings, which were wisely hurrying back to the water, an adult male ran over in an intimidating manner and scattered the young apes. Before leaving the scene, he walked over to the last duckling still on land. With a quick hand movement, like that of a child playing marbles, he flicked it into the moat.

In this case, too, it was as if the ape imagined what might be best for a different organism, obviously having learned to associate ducks with water. I call this *targeted helping,* which is assistance geared toward another's specific situation or need. I believe that apes are masters at this kind of insightful help. Yoni's behavior toward Kohts was anything but exceptional: It is part of the strong sympathetic tendencies of apes recognized by those who work with them. We also don't need to rely on anecdotes, such as Yoni's or Kuni's, since consolation and helping are so common that one can actually measure how apes act around distressed individuals and demonstrate that it's quite different from their usual behavior. By now, consolation is a well-studied phenomenon, as solidly established as aggression or play.

It is unclear how widespread this phenomenon is in other animals, but man's best friend, the dog, may need to be included. There are obviously many anecdotes of people who have received comfort from their dog in times of distress. Take Marley, the Labrador in John Grogan's *Marley & Me,* who was notoriously destructive and boisterous, yet stood perfectly still and silently pressed his head against the belly of Grogan's weeping wife, Jenny, after she had learned about her miscarriage. Charles Darwin relates how a particular dog would never walk by a basket where a sick friend, a cat, lay without giving her a few licks with his tongue. Darwin saw this as a sure sign of the dog's kind feelings.

In the case of canines, too, we don't necessarily need to rely on stories, since there are serious studies. The first one occurred unintentionally, when American psychologist Carolyn Zahn-Waxler sought to determine at what age children begin to comfort family members instructed to sob or cry "Ouch." It turns out that children do so already at one year of age, long before language plays much of a role in their reactions. In the same study, the investigators accidentally discovered that household pets react similarly. Appearing as upset as the children by the distress-faking family members, the pets hovered over them, putting their heads in their laps with what looked like great concern.

But perhaps pets only act like this around humans—who feed and command them—but not with one another? This question was answered by a study modeled after those done on primates, which measured the aftermath of dog fights. Belgian biologists watched close to two thousand spontaneous fights among dogs released every day onto a meadow at a pet food company. After aggressive outbursts, nearby dogs would approach the contestants—most often the losers—to lick, nuzzle, sit together, or play with them. Doing so seemed to settle the group, which quickly resumed its usual activities.

The ancestor of the dog, the wolf, probably behaves the same way. If "man is wolf to man," as Thomas Hobbes liked to say, we should therefore take this in the best possible way, including a tendency to comfort the whimpering victims of aggression.

Changing Places in Fancy

In *Jarhead*, Anthony Swofford describes his time as a U.S. Marine in the Persian Gulf War. The day before they were going to fight an enemy believed to have chemical weapons, one of his buddies, Welty, organized a hugging session:

> We are about to die in combat, so why not get one last hug, one last bit of physical contact. And through the hugs Welty has helped make us human again. He's exposed himself to us, exposed his need, and we in turn have exposed ourselves to him, and for that we are no longer simple grunt savages in the desert ready to jump the Berm and begin killing.

Comforting body contact is part
of our mammalian biology, going
back to maternal nursing, holding,
and carrying, which is why we both
seek and give it under stressful circumstances. People touch and hug
at funerals, in hospitals around sick
or injured loved ones, during wars
and earthquakes, and following defeat in sports. One of the most famous images of a comforting hug
is a grainy black-and-white photograph in which an American soldier

Consolation is a common response to
distress, despair, or grief, such as between
soldiers in the midst of war.

tenderly holds the head of another against his chest. The latter's
friend had just been killed in action during the Korean War.

For his book *Two in Bed,* sociologist Paul Rosenblatt interviewed
couples who had gone through the nightmare of losing a child, noting
how they "quite often would tell me that they dealt with their grief by
holding each other and talking together in bed at night." Given what
contact comfort does for our psychological well-being, one can only
wonder at the "no-hug policy" of a middle school in Virginia. Students could be sent to the principal's office for hugging, holding
hands, or even high-fiving. Trying to stop inappropriate behavior, the
school had come up with a rule that banned the most elementary expressions of affection.

When we comfort others—or for that matter when dogs or
chimps do—what's the motivation behind it? Some of it may be done
to comfort ourselves. Seeing someone cry, we get upset, so by consol-
ing the other, we also reassure ourselves. I am quite familiar with such
behavior in young rhesus monkeys. Once, when an infant was bitten
because it had accidentally landed on a dominant female, it screamed
so incessantly that it was soon surrounded by other infants. I counted
eight of them climbing on top of the poor victim, pushing, pulling,
and shoving one another as well as the first infant. That obviously did

little to alleviate its fright. The monkeys' response seemed automatic, as if they were as distraught as the victim and sought to comfort themselves as much as the other.

This can't be the whole story, though. If these monkeys were just trying to calm themselves, why did they approach the victim? Why didn't they run to their mothers? Why seek out the actual source of distress and not a guaranteed source of comfort? Surely, there was more going on than emotional contagion. The latter can explain a need for comfort, but not the magnetic pull toward a crying peer.

In fact, animals as well as young children often seek out distressed parties without any indication that they know what's going on. They seem blindly attracted, like a moth to a flame. Even though we like to read concern about the other into their behavior, the required understanding may not be there. I will call this blind attraction *preconcern*. It is as if nature has endowed the organism with a simple behavioral rule: "If you feel another's pain, get over there and make contact."

One might counter that such a rule would prompt individuals to waste energy on all sorts of distraught parties, many of whom they'd better stay away from. Approaching others in a predicament may not be the smartest thing to do. But I don't think we need to worry about this, given the evidence that emotions are picked up more readily between parties with close ties than between strangers. A simple approach rule would automatically propel individuals toward those distressed parties that matter most to them, such as offspring and companions.

If true, the sort of behavior that we associate with sympathy arose in fact *before* sympathy itself. If this seems like putting the cart before the horse, it's really not as strange as it sounds. There are other examples of behavior preceding understanding. Language development, for example, doesn't start with children naming things or expressing thoughts. It starts with *babbling*: Babies crawl around uttering nonsensical strings of "ba-ba-ba-ba-ba," advancing to "do-ko-yay-day-bu." When our species claims to be the only talking primate, babbling is obviously not what we have in mind, but this is no reason to belittle it. The fact that everyone's linguistic career starts with this baby lingua franca—which is identical across the globe—illustrates how deeply ingrained language is. It develops out of a primitive urge without any of the refinements of the final product, exactly what I'm proposing for the impulse to attend to someone else's distress.

Preconcern is an attraction toward anyone whose agony affects you. It doesn't require imagining yourself in the other's situation, and indeed the capacity to do so may be wholly absent, such as when a one-year-old child is drawn toward an upset family member. Children of this age are not yet capable of grasping someone else's situation. Preconcern may also explain why certain animals, such as household pets or Oscar the Cat, contact others in pain or near death, or why infant monkeys pile on top of a hapless vocalizing peer.

With preconcern in place, learning and intelligence can begin to add layers of complexity, making the response ever more discerning until full-blown sympathy emerges. Sympathy implies actual concern for the other and an attempt to understand what happened. Yoni's pulling at Kohts's hands comes to mind, as he seemed intent on reading her eyes. The observer tries to figure out the reason for the other's distress, and what might be done about it. Since this is the level of sympathy that we, human adults, are familiar with, we think of it as a single process, as something you either have or lack. But in fact, it consists of many different layers added by evolution over millions of years. Most mammals show some of these—only a few show them all.

Fully developed sympathy is unlikely to be found in rodents, and is probably also absent in canines and monkeys, but some large-brained animals may share the human capacity to put themselves into someone else's shoes. Whether they do or don't has been debated ever since an American primatologist, Emil Menzel, conducted his studies in the 1970s. Do chimps have any inkling of what others feel, want, need, or know? Menzel's groundbreaking work is rarely mentioned anymore, but reading his papers always gives me the feeling of fresh discovery. He was the very first to see the importance of this issue.

Working outdoors in Louisiana with nine juvenile chimps, Menzel would take one of them out into a large, grassy enclosure to reveal hidden food or a scary object, such as a (toy) snake. After this, he would bring this individual back to the waiting group, and release them all together. Would the others appreciate that one among them knew something of importance, and if so, how would they react? Could they tell the difference between the other having seen food or a snake?

They most certainly could. They couldn't wait to follow a chimp who knew a food location but were hesitant to stay close to one who'd seen a hidden snake. This was emotional contagion in action: They copied the other's enthusiasm or alarm. The scenes around food were quite amusing, especially if the "knower" ranked below the "guessers." The below ensued when Belle had been shown food, whereas the alpha male, Rock, had no clue:

> If Rock was not present, Belle invariably led the group to food and nearly everybody got some. In tests conducted when Rock was present, however, Belle became increasingly slower in her approach to the food. The reason was not hard to detect. As soon as Belle uncovered the food, Rock raced over, kicked or bit her, and took it all.
>
> Belle accordingly stopped uncovering the food if Rock was close. She sat on it until Rock left. Rock, however, soon learned this, and when she sat on one place for more than a few seconds, he came over, shoved her aside, searched her sitting place and got the food.

Later on, Belle learned not to approach the food, not even look in its direction, if Rock could see her. She'd sit farther and farther away, or would lay a false track, such as leading Rock to a spot where only one little piece of food had been hidden. She let him have this, while she rushed to the larger pile. The persistence of Rock in following Belle around suggests a conviction on his part that she knew something that she didn't wish to reveal, which is the sort of perspective-taking often referred to as *theory of mind*. Rock seemed to have an idea (a theory) of what might be going on in Belle's head.

The problem with this terminology is that it makes it
sound as if understanding others is an abstract process not unlike the way we figure out how
water turns into ice or why our
ancestors started walking upright. I seriously doubt that we,
or any other animal, can grasp
someone else's mental state at a
theoretical level. All that Rock

Emil Menzel was the first to test what apes know about what others know. One juvenile chimp pokes with a stick at a snake in the grass. From the first chimp's body language, the onlookers know to be cautious.

seemed to be doing was reading Belle's body language and guessing
her intentions. He must have learned that whenever Menzel
showed up, there would be yummy food around that Belle would
try to lay her hands on. So Rock paid close attention to where she
looked, and in which direction she moved; he was acting more like
a hunter than a theoretician.

Menzel's guesser-versus-knower test has inspired a huge following, as reflected in numerous studies on children, apes, birds, dogs,
and so on. This research has taught us that taking someone else's perspective is not limited to human adults. It is best developed in animals
with large brains, but those with smaller brains don't necessarily lack
the capacity. Let me give three typical examples of how this has been
evaluated:

- Human children are champion mind readers. Already at an early
  age, they realize that not everyone knows what they know. In
  one experiment, they watch a character, Maxi, hide chocolate in a
  drawer and then go away. But Maxi's mother accidentally moves
  the chocolate to another place. Where will Maxi be looking
  when he returns? Will it be where the children know the chocolate to be (where the mother put it) or where Maxi last saw it (in
  the drawer)? Most four-year-olds give the right answer, thus taking Maxi's perspective even though they know it to be wrong.

- Indah, a female orangutan at the National Zoo in Washington, D.C., developed the habit of guiding people to food outside her cage. She'd stop a passing caretaker, grasp her, and turn her around so that she'd be facing the dropped food. Then Indah would gently shove her in the food's direction so that the caretaker would pick it up and hand it to her. But what would Indah do with someone unable to see? Let's say, someone with a bucket over her head? Given a choice, Indah would prefer to recruit a seeing experimenter, but working with one with a covered head, she'd first lift the bucket off her head before pushing her toward the food. She developed this clever technique on her own, which was tested further by using a transparent bucket. Since Indah would leave the transparent bucket alone, she seemed to understand that others need sight to be of any assistance.

- Ravens have large brains and are among the smartest birds. Thomas Bugnyar has seen in these birds deceptive tactics reminiscent of Menzel's apes. A low-ranking male was an expert at opening containers but often lost the goodies found inside to a dominant male, who'd steal them from him. The low-ranking male learned to distract his competitor, however, by enthusiastically opening empty containers and acting as if he were eating from them. When the dominant bird found out, "he got very angry, and started throwing things around." Bugnyar further documented that when ravens approach hidden food, they take into account which other ravens may have seen the food being hidden. If their competitor has the same knowledge as they do, a raven will hurry to get there first. With competitors who lack such knowledge, they take their time.

It's not hard to see how much these ingenious experiments owe to Menzel's original study: Treats are being hidden, then being found, and the trick is to know what others know (or, more precisely, what others may have seen). It's ironic that this kind of research grew out of work on primates even though for a while it became fashionable to doubt that nonhuman animals are able to grasp the mental states of others. This doubt has now largely evaporated. Given the latest studies, the line between children and apes has become blurred, as has the line between apes, monkeys, and other animals. Only the most advanced forms of knowing what others know may be limited to our own species.

Yet all the ink spent on this topic can't obscure the fact that we're dealing with a limited phenomenon. I like to call it "cold" perspective-taking, because it focuses entirely on how one individual perceives what another sees or knows. It doesn't concern itself much with what the other wants, needs, or feels. Cold perspective-taking is a great capacity to have, but empathy rests on a different kind, geared more toward the other's situation and emotions. Long ago, Adam Smith aptly described the latter as "changing places in fancy with the sufferer."

We hear the screams of children in the window of a burning house, which alerts us to them and pulls at our heartstrings. But then we look around and evaluate our options. Can they jump? Are we ready to catch? Did someone call the firefighters? Is there a way out of the house, or a way in? It is this combination of emotional arousal, which makes us care, and a cognitive approach, which helps us appraise the situation, that marks empathic perspective-taking. These two sides need to be in balance. If emotions run too high, the perspective-taking may be lost in the process, such as tragically happened at the Singapore Zoo. When a juvenile orangutan got her neck caught in a rope, her mother kept tugging at her to free her. Zoo keepers who tried to interfere were pushed aside by the mother, whose rescue attempt became so frantic that she ended up dislocating her daughter's neck, thus killing her.

Contrast this with a similar situation at a Swedish zoo. A four-year-old juvenile chimpanzee was close to choking, hanging in a climbing rope, with the rope wrapped twice around his neck. He struggled silently, his feet dangling. The oldest, most dominant male of the group went over to him, picked up the victim with one arm, thus relieving the tension on the rope, and unwrapped the rope with his free hand. He then carried the juvenile to the ground and gently put him down. It was all over in seconds, with just a few quick hand movements. The only sound came from a screaming caretaker.

Perhaps the orangutan mother's urge to save her daughter had been too intense for her to think clearly. Or perhaps she lacked experience with ropes. The male chimpanzee, in contrast, stayed calm and did the right thing. It takes great intelligence to inhibit the most natural impulse, which is pulling, and replace it with a more effective course of action. Such cases illustrate the two-tiered process underlying helping: emotion and understanding. Only when both processes are combined can an organism move from preconcern to actual concern, including the targeted helping typical of our close relatives.

Jumping into Water

Atlanta is the Mecca of primatology. Menzel's son, Charles, who followed in his father's footsteps studying chimpanzees, lives just a few blocks from my home in Stone Mountain. One day, when grandpa had come to see his grandchildren, I snared Emil for an interview in my kitchen over a cup of soup. He was in his early seventies.

Although born and raised in India, Emil is a typical southern gentleman: mild-mannered, courteous, and with a great sense of humor. He is still very much committed to his pioneering ideas, which concur with mine, in that he has a high opinion of ape intelligence and thinks that the main factor limiting scientific discovery is human imagination and creativity, not the ability or inability of apes to meet expectations.

He told me about the publication of his hide-and-seek study, and how he wanted to move on to other questions but kept being invited to lecture on this particular experiment. It obviously had struck a chord. One of the invitations was to an East Coast college at which an eminent behaviorist chaired the session and annoyed Menzel. First of all, he didn't give the audience any chance to speak, and second, he lectured the speaker, saying that, since chimps are hard to deal with, it would be far more practical to work with pigeons. This related to the curious opinion at the time that it doesn't matter which animal one studies: Since all animals rely on stimulus-response learning, a chimpanzee really isn't doing anything different from a pigeon.

The professor, however, walked into a trap of sorts, because Menzel had decided to show a spectacular escape that he had filmed a few years before. His chimps had put a long pole against the wall of their enclosure, with some individuals holding the pole steady while others scaled it to get out. Not the sort of thing pigeons do every day. Menzel had decided to accompany his movie with as neutral a narrative as possible, avoiding reference to complex mental operations. He'd just say matter-of-factly, "See how Rock grabs the pole while glancing at the others," or "Here a chimpanzee swings over the wall."

After Menzel's talk, the eminent professor stood up to accuse him of being unscientific and anthropomorphic, of attributing plans and intentions to animals that obviously didn't have either. To a roar of approval, Menzel countered that he had not attributed anything, that if this professor had seen plans and intentions he must have seen them with his own eyes, because Menzel himself had refrained from alluding to any such things.

It's impossible to watch chimps and not notice their smarts. Menzel told me that he sometimes speculates on how much more evidence there might be hidden in the many pages of handwritten notes he took during his experiments (I suggested, of course, that now that he's retired, nothing should keep him from going back to those notes, but his shrug told me not to hold my breath). He strongly believed in watching over and over and thinking through what his observations might mean, even if he'd seen a certain behavior only once. He objected to calling single observations mere "anecdotes," adding with a mischievous smile, "My definition of an anecdote is somebody else's observation." If you have seen something yourself, and followed the entire dynamic, there usually is no doubt in your mind of what to make of it. But others may be skeptical and need convincing.

This is an important point, because the most striking examples of empathic perspective-taking, both in humans and animals, concern single incidents. One day, such an incident happened while I stood admiring water lilies in a large pond in Balboa Park, San Diego. The pond lacked any protection and was next to a path with people. A small child, perhaps three years old, raced through the crowd straight into the pond. I was surprised how quickly he sank: One moment we heard the splash, the next he was gone. But before anyone could take action, his mother jumped in, too. She emerged soon with her son in her arms. She must have been running after him, knowing what might happen, and without any hesitation, obviously fully dressed, followed him into the pond. If she hadn't done so, who knows how long it might have taken to find the child, as the water was very murky.

Here we see alertness to another's situation at a level that's impossible to mimic in the lab. We can ask people what they will do under certain circumstances, and we can test them under mildly upsetting conditions, but no one is going to reenact the near drowning of a child to see how its parents react. Yet this sort of situation, which is essentially untestable, produces by far the most interesting altruism and the most relevant to survival. The same applies to animals: We can explore how they react to hidden objects, we may even test their perception of distress calls, but who will set up experiments in which a friend or relative is being strangled by a rope around their neck? Not me, and not most other scientists. All we can go by is the occasional report of how apes respond to such calamities.

For humans, anecdotes appear in the daily newspaper. For example, New Yorkers who escaped from the World Trade Center on 9/11 describe the brave firefighters who walked past them, burdened with lifesaving gear. The firefighters were going up as the people were coming down. People were beginning to panic, but the firefighters acted with great confidence, telling them to get out of the building, thus assuring an orderly exit while they themselves were walking toward their death.

Or take Army Sergeant Tommy Rieman, who came under fire

during a 2003 ambush in Iraq. Shielding his gunner from attack with his own body, he began returning fire. He suffered several bullet and shrapnel wounds, yet refused medical attention until the wounded had been extracted from the scene. Every natural calamity produces its heroes, who run into burning houses or dive into icy rivers to save total strangers. During the German occupation of Europe, scores of people risked their lives shielding Jews, such as Anne Frank's family in Amsterdam. During famines, farmers commonly share precious food with hungry city-dwellers. The 2008 earthquake in central China even produced a national "Mother Number One," a policewoman who breastfed a number of orphaned babies on the scene. The mother of an infant herself, Jiang Xiaojuan felt that she had milk to spare.

None of this would happen without our capacity for empathy. There are, in fact, so many stories of human self-sacrifice that we rightly consider such behavior a characteristic of our species and are keen to recognize it in our ancestors. When the fossil of a completely toothless hominid was recently found in the Caucasus, scientists suggested that survival would have been impossible for this individual without extensive feeding and care. They concluded that these forebears must have been humanlike, even though they lived almost two million years ago, since they practiced compassion.

But this assumes that compassion is restricted to our lineage. Some animals, too, feed those who have trouble feeding themselves. For example, in Gombe National Park in Tanzania, an old and sick female chimpanzee, named Madame Bee, had trouble climbing fruit trees, and sometimes depended on her daughters:

> She looked up at her daughters, then lay on the ground and watched as they moved about, searching for ripe fruits. After about ten minutes Little Bee climbed down. She carried one of the fruits by its stem in her mouth and had a second in one hand. As she reached the ground, Madame Bee gave a few soft grunts. Little Bee approached, also grunting, and placed the fruit from her hand

on the ground beside her mother. She then sat nearby as the two
females ate together.

There is in fact so much evidence for altruism in apes that I will
pick just a handful of stories to drive home my point. Some incidents,
such as the one above, concern related individuals, but similar behavior occurs between unrelated ones. At our primate center, we have an
old female, Peony, who spends her days with other chimpanzees in a
large outdoor enclosure. On bad days, when her arthritis is flaring up,
she has great trouble walking and climbing. But other females help
her out. For example, Peony is huffing and puffing to get up into the
climbing frame in which several apes have gathered for a grooming
session. An unrelated younger female moves behind her, places both
hands on her ample behind, and pushes her up with quite a bit of effort, until Peony has joined the rest.

We have also seen Peony get up and very slowly move toward the
water spigot, which is at quite a distance. Younger females would
sometimes run ahead of her, take in some water, then return to Peony
and give it to her. At first we had no idea what was going on, since all
we saw was one female placing her mouth close to Peony's, but after a
while the pattern became clear: Peony would open her mouth wide
and the younger female would spit a jet of water into it.

Chimp Haven, an organization dedicated to the retirement of
laboratory chimpanzees onto large forested islands, adopts many
chimps without outdoor experience and hence with no knowledge of
grass, bushes, and trees. One naïve female, Sheila, had formed a bond
with an unrelated younger one, Sara, who did know trees and was not
afraid to climb into them. Before Sheila learned to do the same from
watching her friend, Sara would occasionally break out a leafy branch
that she'd bring down specifically for Sheila to nibble on.

Sara also once saved Sheila from a snake. Sara saw the snake first
and sounded the alarm by barking loudly, whereupon Sheila approached to take a look. Sara held her friend back by her arm, vigorously pushing her backward. Sara was poking the snake with a stick

and getting closer while continuing to hold Sheila back. It was later determined that the snake was poisonous.

One might argue that none of these acts compares to running into a burning building: None are terribly risky or costly. In fact, I once heard an eminent psychologist tell an audience that altruism could indeed be found in other animals, but that they invariably gave priority to their own survival. "An ape will never jump into a lake to save another," he declared with great aplomb. As soon as these words had left his mouth, however, I began racking my brain for contradictory information, sure that I'd heard otherwise. Apes and water are in fact a dangerous mix, much more so than humans and water, because apes can't swim. Chimpanzees have been known to panic and drown in knee-deep water. They sometimes learn to overcome this fear, but for an ape to enter water takes extraordinary courage.

Zoos often keep apes on islands surrounded by water-filled moats, and there indeed exist reports of them trying to save companions, sometimes with a fatal outcome for both rescuers and victims. One male lost his life when he waded into water to reach an infant who had been dropped by an incompetent mother. At another zoo, an infant chimpanzee hit an electric wire and panicked, jumping off his mother into the water, whereupon mother and son drowned together when she tried to save him. And when Washoe, the world's first language-trained chimp, heard another female scream and hit the water, she raced across two electric wires that normally contained the apes to reach the victim, who was wildly thrashing about. Washoe waded into the slippery mud at the edge of the moat to grab one of the female's flailing arms, and pulled her to safety.

Obviously, hydrophobia cannot be overcome without an overwhelming motivation. Explanations in terms of mental calculations ("If I help her now, she will help me in the future") don't cut it: Why would anyone risk life and limb for such a shaky prediction? Only immediate emotions can make one abandon all caution. Such heroism is common in chimpanzee social life. For example, when a female reacts to the screams of her associate by defending her against a dominant male, she endangers herself on behalf of another. I have often seen female chimps take a serious beating for their friends. In the wild, even more risky rescues have been observed when chimpanzees rally in response to the screams of one among them to a leopard attack. In dense forest, the others usually can't see what's happening, but screams come in many intensities, and the apes recognize the extreme alarm in the victim's voice. The forest immediately fills with angry calls and barks by all chimps within earshot, who quickly converge on the danger and rout the leopard, who is eager to escape such a mob.

Commitment to others, emotional sensitivity to their situation, and understanding what kind of help might be effective is such a human combination that we often refer to it as being *humane*. I do believe that our species is special in the degree to which it puts itself into another's shoes. We grasp how others feel and what they might need more fully than any other animal. Yet our species is not the first or only one to help others insightfully. Behaviorally speaking, the difference between a human and an ape jumping into water to save another isn't that great. Motivationally speaking, the difference can't be that great, either.

Little Red Riding Hood

How foolish of Little Red Riding Hood to think she was visiting Grandma! As every child knows, the bed was occupied by a big bad wolf.

But does every child understand that Little Red Riding Hood was in fact unafraid? Obviously, if the girl knew what we know, she should be very afraid. But since she was ignorant, what was there to worry about? Asked about her state of mind, the correct answer is that she had no fear. Most children, however, give the wrong answer: They can't help but project their own anxiety onto the story's character.

Psychologists count this as a failure: It shows an inability to take someone else's perspective. But I see it differently. Children do in fact take Little Red Riding Hood's viewpoint in a way that suits an emotionally charged situation. They put themselves in her place, imagining themselves standing in front of Grandma's bed with their basket, but armed with their own knowledge. Naturally, they're scared to death. Psychologists may want a rational evaluation, but children have a hard time extracting themselves from a confrontation with a salivating predator. Only by the age of seven or eight do they manage such distance—and we applaud them for understanding that Little Red Riding Hood actually isn't afraid—but the real lesson here is the overwhelming power of emotional identification.

Instead of staying neutral, children tend toward empathy. This primal connection automatically takes over if anyone they feel close to gets into trouble, and it applies equally to adults. Horror movies play to this tendency. They hit us below the belt, so to speak, relying on a far more visceral identification with the onscreen characters than, say, an Ingmar Bergman movie does. When our favorite character approaches the ax murderer hiding behind the shower curtain, we don't worry too much about what she knows or doesn't know.

The child's capacity to emotionally enter another's shoes and guess what he or she feels has been tested. For example, a child watches an adult open a gift box. The child is not allowed to peek inside, but if the person happily exclaims "Oh boy!" the child guesses that there must be something good inside, such as candies. If the experimenter looks disappointed, on the other hand, saying "Oh no!" the child understands that the box must contain something distasteful, like broccoli. Their reaction is not that different from Menzel's apes, who recognized if one among them had spotted hidden food or danger.

Children read "hearts" well before they read minds. At a very young age, they already understand that other people have wants and needs, and that not everybody necessarily wants or needs the same. They recognize, for example, that a child looking for his rabbit will be happy to find it, whereas a child searching for his dog will be largely indifferent to finding a rabbit.

We take such abilities for granted, but have you ever noticed that not everyone takes advantage of them? I'm talking about adults here, such as the two kinds of gift givers we're all familiar with. Some friends will go out of their way to find you a gift that *you* might like. Knowing that I love opera or play the amateur baker at home, they buy me a CD of the latest Anna Netrebko performance or the best rye flour in town. I always feel that the amount of money spent is secondary to the thought, and these people are clearly intent on pleasing me. The other kind of gift giver arrives with what *they* like. They've never noticed that we don't have a single blue item in the house, but since they love blue, they bestow an expensive blue vase upon us. People who fail to look beyond their own preferences ignore millions of years of evolution that have pushed our species to ever better perspective-taking.

Every day, humans are prepared to improve the lives of others, including complete strangers, provided it isn't too much trouble. Strictly speaking, this isn't altruism, because altruism requires an effort. No, I am talking here of a situation that doesn't set you back one bit. An example is what happened during a hike my wife and I once took in Canada. This was during our early days in North America, when every distance seemed ten times longer than we'd ever imagined. We were trying to escape from a lakeshore where giant mosquitoes were eating us alive and had decided to walk to the nearest town. We walked and walked over a never-ending dirt road under a bright sun. A large station wagon with a Canadian family slowed down next to us and the driver nonchalantly leaned out, asking "Do you need a ride?" When he told us how far the town still was, we were more than happy to accept. I still feel grateful.

Low-cost assistance is common in humans, such as one tennis player helping up another.

This is so-called low-cost altruism, when one isn't going

much out of the way for someone else but still offers substantial help. We do it all the time. If someone at the airport drops his boarding pass and I alert him to it, it costs me very little, but saves my fellow passenger much grief. We also customarily hold the door that we just went through for someone who comes after us, slide aside on a park bench for someone who wants to sit, hold back an unknown child who's about to run onto the street at the wrong moment, or help an older person lift a heavy piece of luggage. Humans are great at this sort of assistance, at least under relatively comfortable circumstances, because the behavior vanishes as soon as the *Titanic* starts foundering. Under hardship, the cost of civility goes up.

To be considerate, even in small ways, one needs empathic perspective-taking. One needs to understand the effect of one's behavior on others. As I search for possible animal parallels, a curious behavior comes to mind that I saw among wild chimpanzees while standing in an almost dry river bed in the Mahale Mountains in Tanzania. The chimps were relaxing on large boulders, grooming one another. I had read about their so-called *social scratch*, but never seen it firsthand.

Social scratching occurs when one ape walks up to another, vigorously scratches the other's back a few times with his fingernails, then settles down to groom the other. More back-scratching may follow during the grooming session. The behavior itself cannot be hard to learn for an animal that commonly scratches itself, but here's the rub: When one scratches oneself, this is usually in order to relieve itching (try not to scratch yourself for an hour, and you'll appreciate its importance). But scratching someone else's back is something else entirely: It doesn't do any good for the scratcher himself.

Unlike grooming, the social scratch is unlikely to be innate. We know this, because curiously only the Mahale chimps show this behavior. It hasn't been documented in any other chimpanzee community. Anthropologists and primatologists call such group-specific behavior a "custom." Customs are habits that are passed on within a community and are unique to that community. Eating with knife and

How did the Mahale chimpanzees develop an other-serving custom? The middle individual scratches another's back with long strokes.

fork is a human custom in the West, and eating with chopsticks a custom in the East. By itself, finding customs in chimpanzees is not that special, because these animals have lots of them, more than any species apart from ourselves. The real puzzle is how members of the Mahale community came to adopt a custom that favors others more than themselves.

How do we learn to hold a door open for others? You might say that we have been told to do so by our parents, which is undoubtedly true, but later on such habits are reinforced by experiencing them and appreciating the favor. From this we figure it might be nice to do the same for others. Could this be how the social scratch spread among the Mahale chimps? Imagine that one ape was accidentally scratched by another, and it felt so good that he decided to offer the same experience to a third, perhaps one whom he wanted to ingratiate himself with, such as the boss. This is entirely possible, but would imply perspective-taking. The scratcher would need to translate a bodily experience into an action that re-creates the same experience in somebody else. He'd need to realize that others feel what he feels.

The social scratch is a deceptively simple act, behind which dwells a profound mystery that can't be resolved by observations alone. I could watch as many of these interactions as Toshisada Nishida, my host in Mahale, has seen in his four decades in the field, and would still have no clue what's behind them. We cannot ask the chimps why they do it, and we are too late to witness the first social scratch, the one that seeded the custom. This is where research in captivity offers a solution: Problems from the field can be taken into a setting that allows systematic testing. We can see, for example, how sensitive primates are to another's welfare if we give them the opportunity to do small favors.

Over the last few years, interest in this question has grown. Let me start with two simple studies with our own capuchin monkeys. I have two groups of these cute brown monkeys. They have outdoor space, where they can sit in the sun, catch insects, groom, and play. There's also an indoor area with doors and tunnels that make it easy to move them into tests. They are used to the procedures, and actually eager to be tested, which almost always involves attractive food. The capuchin is a favorite primate for these kinds of experiments, because they are extremely smart (they have the largest brain relative to body size of all monkeys), share food, and cooperate easily with one another as well as with humans. They are such appealing monkeys that my students have pictures of their darlings on the wall and passionately talk about them as if they're discussing a soap opera.

Our first experiment tested whether these monkeys recognize the needs of others. Do they understand when one among them is hungry? They indeed seem to do so, because we found that their willingness to share food with another depended on whether they had seen the other just eat. They shared more with a monkey who'd been empty-handed than one whom they'd seen munching on food.

The second experiment was even more revealing since it suggested interest in another's welfare. We placed two monkeys side by side: separate, but in full view. One of them needed to barter with us, which is something these monkeys understand naturally. For example, if we leave a broom behind in their enclosure, all we need to do is point at it and hold up a peanut, and the monkeys understand the deal: They bring us the broom in exchange. In the experiment, the bartering was done with small plastic tokens, which we'd first give to a monkey, after which we'd hold out an open hand, letting them return the token for a tidbit.

The interesting test came when we offered a choice between two differently colored tokens with different meaning: One token was "selfish," the other "prosocial." If the bartering monkey picked the selfish token, it received a small piece of apple for returning it, but its partner got nothing. The prosocial token, on the other hand, rewarded both monkeys equally at the same time. Since the monkey who did the bartering was rewarded either way, the only difference was in what the partner received. To make sure they understood, Kristi Leimgruber, my assistant, would make quite a show by raising either one hand with food and feeding one monkey, or raising both hands and simultaneously handing food to both of them.

We know exactly how socially close any two monkeys are because we watch how much time they spend together in the group. We found that the stronger the tie with its partner, the more a monkey would pick the prosocial token. The procedures were repeated many times with different combinations of monkeys and different sets of tokens, and they kept doing it. Their choices could not be explained by fear of punishment, because in every pair the dominant monkey (the one who had least to fear) proved the more prosocial one.

Does this mean that capuchin monkeys care about the welfare of others? Do they like to do them favors? Or could it be that they just love to eat together? If both monkeys are rewarded, they will sit side by side munching on the same food. Do things taste better together than alone, the way we are more at ease having dinner with family and friends? Whatever the explanation, we showed that monkeys favor sharing over solitary consumption.

Similar experiments with apes initially failed, leading to premature headlines in the media such as "Chimpanzees Are Indifferent to the Welfare of Unrelated Group Members." But as the old saying goes, absence of evidence is not evidence of absence. All that we seem to have learned from these experiments is that humans can create situations in which apes put their own interests first. With regard to our own species, too, this wouldn't be hard to do. Take the way people trample one another to get to the merchandise when a department store opens its doors for a major sale. In 2008, a store employee was killed in the process. But would anyone conclude from these scenes that humans, as a species, are indifferent to one another's welfare?

Successful approaches often require a flash of insight into what best suits a particular animal. Once achieved, the false negatives will be forgotten. This is what happened when Felix Warneken and colleagues of the Max Planck Institute in Leipzig, Germany, hit on a winning formula to test ape altruism. They worked with chimpanzees at a sanctuary in Uganda, where the apes spent their days on a large, lush island with lots of trees. Every night they were brought inside a building, which is where the tests took place. A chimp would watch a human unsuccessfully reach through the bars for a plastic stick. The human would not give up, but the stick would stay out of reach. The chimp, however, was in an area where he could just walk up to the stick. Spontaneously, the apes would help the reaching person by picking up the desired item and handing it to him. They were not trained to do so, and rewarding them for their effort made no difference. A similar test with young children led to the same outcome.

When the investigators increased the cost of helping, by having the apes climb up a platform to retrieve the stick, they still did so. The children also helped even if obstacles were put in their way. Obviously, both apes and children spontaneously help others in need.

But could it be that chimps in a sanctuary help humans because their lives depend on them? Prepared for this argument, the investigators had selected human partners who were barely known to the apes, and certainly not involved in their daily care. They further added a second test to see if the apes would assist one another in the same way.

From behind bars, one chimpanzee would watch a partner struggle to open a door leading to a room where both knew there was food. The room was closed, however. The only way to get in would be if a chain blocking the door were removed, but this chain was beyond the control of the partner. Only the first chimp could undo it. The outcome of this particular experiment surprised even me—I wasn't sure what to predict given that all the food would go to the partner. Yet the results were unequivocal: One chimp removed the peg that held the chain, thus allowing its companion to reach the food.

What these apes did was far more complex than the choice between tokens faced by our monkeys. They needed to understand the other's intentions and decide on the best solution for what the other wanted. They showed targeted helping, just as apes do in everyday life. The basic motivation behind their assistance, however, was probably not so different from that of the monkeys: Both care about the well-being of those around them. Traditional views based on payoffs for the actor just can't explain these results: For the monkeys prosocial choices didn't yield any more rewards than selfish ones, and for the chimps, too, rewards made no difference.

Warm Glow

Perhaps it is time to abandon the idea that individuals faced with others in need decide whether to help, or not, by mentally tallying up costs and benefits. These calculations have likely been made for them by natural selection. Weighing the consequences of behavior over evolutionary time, it has endowed primates with empathy, which ensures that they help others under the right circumstances. The fact that empathy is most easily aroused by familiar partners guarantees that assistance flows chiefly toward those close to the actor. Occasionally, it may be applied outside this inner circle, such as when apes help ducklings or humans, but generally primate psychology has been designed to care about the welfare of family, friends, and partners.

Humans are empathic with partners in a cooperative setting, but "counterempathic" with competitors. Treated with hostility, we show the opposite of empathy. Instead of smiling when the other smiles, we grimace as if the other's pleasure disturbs us. When the other shows signs of distress, on the other hand, we smile, as if we enjoy their pain. One study described reactions to a hostile experimenter as follows: "His euphoria produced dysphoria and his dysphoria produced euphoria."

So, human empathy can be turned into something rather unattractive if the other's welfare is *not* in our interest. Our reactions are far from indiscriminate, exactly as one would expect if our psychology evolved to promote within-group cooperation. We are biased toward those with whom we have, or expect to have, a positive partnership. This unconscious bias replaces the calculations often as-sumed behind helping behavior. It's not that we are incapable of calculations—we do sometimes help others based purely on ex-pected returns, such as in business dealings—but most of the time human altruism, just like primate altruism, is emotionally driven.

When a tsunami hits people a world away, what makes us decide to send money, food, or clothes? A simple newspaper headline "Tsunami in Thailand Kills Thousands" won't do the trick. No, we re-spond to the televised images of dead bodies on the beach, of lost chil-dren, of interviews with tearful victims who never found their loved ones. Our charity is a product of emotional identification rather than rational choice. Why did Sweden, for example, offer such massive support to the affected region, making a substantially larger contribu-tion than other nations? More than five hundred Swedish tourists lost their lives in the 2004 disaster, a fact that aroused great solidarity in Sweden with the affected people in Southeast Asia.

But is this altruism? If helping is based on what we feel, or how we connect with the victim, doesn't it boil down to helping ourselves? If we feel a "warm glow," a pleasurable feeling, at improving the plight of others, doesn't this in fact make our assistance selfish? The problem is that if we call this "selfish," then literally everything becomes selfish, and the word loses its meaning. A truly selfish individual would have no trouble walking away from another in need. If someone is drown-ing: Let him drown. If someone is crying: Let her cry. If someone drops his boarding pass: Look away. These are what I'd call selfish reactions, which are quite the opposite of empathic engagement. Empathy hooks us into the other's situation. Yes, we derive pleasure from help-ing others, but since this pleasure reaches us *via* the other, and *only* via the other, it is genuinely other-oriented.

At the same time, there is no good answer to the eternal question of how altruistic is altruism if mirror neurons erase the distinction between self and other, and if empathy dissolves the boundaries be-tween people. If part of the other resides within us, if we feel one with the other, then improving their life automatically resonates within us. And this may not be true only for us. It's hard to see why a monkey would systematically prefer prosocial over selfish outcomes if there weren't something intrinsically rewarding about the former.

Perhaps they too feel good doing good.

5

The Elephant in the Room

Seeing himself in the mirror for the first time, the chimpanzee opened his mouth in amazement and looked questioningly and curiously at the glass, as though asking silently but eloquently: "Whose is this face over there?"

—NADIA KOHTS, 1935

Y ou'd think you'd hear an elephant approach. But you can stand sweating in the sun in a forest clearing in Thailand while one of them comes up from behind, and you won't feel any vibrations, won't hear a thing, because elephants are perfectly elastic, walking on velvet cushions while carefully avoiding branches or leaves that might snap under their feet. They're in fact remarkably elegant animals.

They're also dangerous. The U.S. Bureau of Labor Statistics rates elephant keeper as the single most dangerous profession. In Thailand alone, more than fifty mahouts (caretakers/trainers) are killed every year. One problem is the unexpected speed of these animals. Another is their cuddly "jumbo" image, which pulls us toward them and makes us lower our guard. The appeal that elephants hold for humans is nothing less than astonishing, and was already witnessed in ancient Rome, not a place known for squeamishness. Pliny the Elder describes the way the crowd reacted to twenty elephants being savaged in an arena:

> ... when they had lost all hope of escape [they] tried to gain the compassion of the crowd by indescribable gestures of entreaty, deploring their fate with a sort of wailing, so much to the distress of the public that they forgot the general and his munificence carefully devised for their honor, and bursting into tears rose in a body and invoked curses on the head of Pompey.

With an anatomy so different from ours, the ease with which elephants arouse human sympathy poses yet another version of the correspondence problem. How do we map their bodies onto ours? We recognize their hostile trunk movements, which are stiff and agitated, but also their gentle rubbing against one another, with trunks feeling into the others' mouths—a most vulnerable place for this organ to visit. Most of all, we recognize their fun when, for example, they jostle in a water hole, completely covered with mud, pushing one another aside until they slide and splash with their eyes turned outward so that we see a lot of white, which gives the impression that they're going crazy. They seem to have a sense of humor.

I had come to North Thailand to visit a student, Joshua Plotnik, who's studying social behavior at the Elephant Nature Park, near Chiang Mai, as well as the Thai Elephant Conservation Center, near Lampang. I had seen African elephants on the savanna, but the big difference with these Asian elephants was that I was *not* sitting in a Jeep: I stood there right next to these mighty beasts with their shrill trumpeting sounds and deep, drawn-out rumbles, and sensed right away how tiny and vulnerable the human race is.

Elephants are magnificent. But the elephant situation in Thailand is also a sad one of changing habits. Thousands of elephants used to be employed for timber harvesting, but a devastating flood blamed on deforestation led to a nationwide logging ban in 1989. This created an urgent need to care for the animals, for which the owners lacked income. On top of this, there are the three-legged land mine victims from the Thai-Burma border, and other animals in urgent need of care. Today many elephants serve to educate the public, each one controlled day and night by a personal mahout. That is the only way to take care of these animals short of releasing them. The latter may seem preferable, but in a populated nation such as Thailand, and given the danger elephants pose, "liberation" would mean almost certain death.

It's a bit as if you have a tractor in your garage, which can start its engine on its own anytime and drive out on the road while leveling small dwellings, crushing people, and uprooting leafy vegetation. No one would like to have such a liability, and an elephant in an urban setting would barely be any different. So, under control they are and need to be. I was thoroughly impressed by the commitment of those who maintain them in the parks. The elephants either move together under semifree conditions or conduct shows and trainings, including "orchestral" performances on xylophones and reenactments of their species' historic employment in the timber industry. These performances ensure their upkeep in parks and sanctuaries, some of which let ecotourists pay for the privilege of shoveling dung.

Now, what other animal could generate such devotion?

Ontogeny and Phylogeny

Two tall adolescent bulls at the Elephant Conservation Center effortlessly pick up a long, heavy log with their tusks, each standing on one end, draping their trunks over the log to keep it from rolling off. Then they walk in perfect unison with the log between them, while the two mahouts on their heads sit chatting and laughing and looking around, and are certainly not directing every move. Training is obviously part of this picture, but one cannot train any animal to be so coordinated. One can train dolphins to jump synchronously because they do so in the wild, and one can teach horses to run together at the same pace because wild horses do the same. For the same reason, one can train two elephants to pick up a log and carry it together to another place, walking in the same rhythm, and lowering the log together to set it down on a pile without a sound, because elephants are extraordinarily well coordinated in the wild. They're obviously not picking up logs, but they perform concerted actions to support a wounded companion or calf in need.

I ran into a different kind of cooperation in the Elephant Nature Park, where a blind elephant walked around with her seeing friend. The two females were unrelated yet seemed to be joined at the hip. The blind one was clearly dependent on the other, who seemed to understand this. As soon as the latter moved away, one could hear deep sounds coming from both of them, sometimes even trumpeting, which indicated the other's whereabouts to the blind elephant. This noisy spectacle would continue until they were reunited again. An intensive greeting followed, with lots of ear-flapping, touching, and mutual smelling.

The whole world assumes that these animals are highly intelligent, but in fact there is little official proof. The sort of experiments conducted with monkeys and apes, which reveal what these animals understand, are rarely done with elephants, for the simple reason that they're not easy to work with. Which university is ever going to set up an elephant lab? Anyone who wishes to test elephants will need to either work in countries with a history of controlling them, such as Thailand or India, or in zoo settings. Before he went to Thailand, Joshua worked at the Bronx Zoo, in New York, where he was involved in our first elephant experiment involving a huge mirror.

This experiment sprang from our interest in empathy. Advanced empathy is unthinkable without a sense of self, which is what mirror tests get at. Of all animals, elephants are perhaps the most empathic, so we were curious to see if they had enough self-identity to recognize their reflection. This capacity was predicted decades ago by Gordon Gallup, the psychologist who first showed that apes (but not monkeys) recognize themselves in a mirror.

If I walk up to my capuchin monkeys while wearing sunglasses, some will threaten me as if they don't recognize me, but soon they'll switch to curiosity. They never use their reflection in my glasses to inspect their own bodies, however. They simply don't "get" what they are looking at. How different from the apes, which as soon as they spot my sunglasses begin to make weird grimaces while staring into them. They're never confused about who I am (I could literally arrive in drag, and they'd still know whom they're dealing with), but impatiently jerk their heads at me until I take off the glasses and hold them closer to them as little mirrors. Females then turn around to look at their behinds—a critical part of their anatomy that they never get to see—and open their mouth to inspect the inside, picking at their teeth. Anyone who has seen an ape do so realizes that the animal is not accidentally opening its mouth or turning around: Its eyes monitor its every move in the mirror.

Any large-brained animal with well-developed empathy should be able to do the same, Gallup believed. But why bring up empathy? What, if anything, do mirrors tell us about social skills? Part of the answer can be found in child development. Human babies don't recognize themselves in a mirror right away. A one-year-old is as confused as many animals about the "other" in the mirror, often smiling at, patting, even kissing their reflection. They usually pass the so-called *rouge test* in front of the mirror by age two, rubbing off a small

dab of makeup that has been put on their face. They don't know about the dab until they look into the mirror, so when they touch it we can be sure they connect their reflection with themselves.

Around the same time children pass the rouge test, they become sensitive to how others look at them,

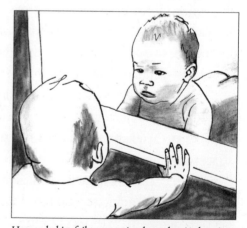

Human babies fail to recognize themselves in the mirror until about eighteen to twenty-four months of age.

show embarrassment, use personal pronouns ("That's mine!" or "Look at me!"), and develop pretend play in which they act out little scenarios with toys and dolls. These developments are linked. Children passing the rouge test use more "I" and "me" and show more pretend play than children failing this test.

I must admit that I find reactions to mirrors per se a fairly boring topic. They don't relate to survival and barely play a role in nature. Plenty of animals that fail to recognize their own reflection are doing perfectly fine. What makes mirror tests exciting is what they tell us about how an individual positions itself in the world. A strong sense of self allows it to treat another's situation as separate from itself, such as when a child first drinks from a glass of water and then offers the same glass to her doll. The child knows very well that dolls don't drink, but nevertheless likes to attribute emotional states to it. The doll is at the same time similar ("like me") and different. The child is turning into a role-player, and finds a thirsty, sad, or sleepy doll the perfect partner since it never objects to her fantasies.

Since all of these abilities emerge at the same time as mirror self-recognition, I'll speak of the *co-emergence hypothesis*. Advanced empathy belongs to the same package. This has been extensively tested on Swiss children by Doris Bischof-Köhler. She'd have a child eat quark (a dairy cream) sitting next to an adult instructed to look sad at a given moment because her spoon broke. The child might pick up an extra spoon left on the table or offer its own spoon. Some children would try to feed their partner with the broken spoon. Or the adult would "accidentally" rip off a limb from her teddy bear, whereupon she was instructed to sob for minutes. Children who repaired the teddy, offered another toy, or stayed close and made eye contact were considered prosocial. When the same children were tested with mirrors, the outcome was entirely in line with the co-emergence hypothesis. Children who had acted prosocially passed the mirror test, whereas those who had given no assistance failed the test.

Why should caring for others begin with the self? There is an abundance of rather vague ideas about this issue, which I am sure

neuroscience will one day resolve. Let me offer my own "hand waving" explanation by saying that advanced empathy requires both
mental mirroring and mental separation. The mirroring allows the
sight of another person in a particular emotional state to induce a
similar state in us. We literally feel their pain, loss, delight, disgust,
etc., through so-called shared representations. Neuroimaging shows
that our brains are similarly activated as those of people we identify
with. This is an ancient mechanism: It is automatic, starts early in life,
and probably characterizes all mammals. But we go beyond this, and
this is where mental separation comes in. We parse our own state
from the other's. Otherwise, we would be like the toddler who cries
when she hears another cry but fails to distinguish her own distress
from the other's. How could she care for the other if she can't even tell
where her feelings are coming from? In the words of psychologist
Daniel Goleman, "Self-absorption kills empathy." The child needs to
disentangle herself from the other so as to pinpoint the actual source
of her feelings.

Note that I am not talking here about self-reflection or introspec-
tion, mainly because we have no way of knowing if preverbal children
or animals possess this kind of self-awareness. Even for our own
species, I am not as convinced as some scientists are that the questions
humans answer about themselves reveal what they truly experience.
More interesting than self-reflection is the self-other boundary. Do we
see ourselves as a separate entity? Without a concept of self, we'd lack
mooring. We'd be like little boats floating and sinking together. One
wave of emotion, and we'd move up or down with it. In order to show
genuine interest in someone else, offering help when required, one
needs to be able to keep one's own boat steady. The sense of self serves
as anchor.

Long ago, before all of the above was known, Gallup proposed
that the way a given species processes mirror information tells us
something about this sense of self. Certain cognitive capacities can be
expected only in species that recognize themselves, he said. Since this
idea resembled what happens during child development, it reminds

me of a classic book by Stephen Jay Gould, *Ontogeny and Phylogeny*, which discusses comparisons between individual development (ontogeny) and the evolution of species (phylogeny). Both processes concern vastly different time scales yet show striking parallels. Similarly, the co-emergence hypothesis postulates parallels between ontogeny and phylogeny in that the same capacities that develop together in a two-year-old child evolved together in some animal species.

If so, species that recognize themselves in a mirror should be marked by advanced empathy, such as perspective-taking and targeted helping. Species that don't recognize themselves, on the other hand, should lack these capacities. This is a testable idea, and Gallup felt that the prime candidates to look at, apart from the apes, would be dolphins and elephants.

Dolphins were the first to fit his prediction.

Flippin' Idiots

No one ever blinks an eye if a pop star is called "empty-headed," or an unpopular American president a "chimp"—even though the primatologist winces at the latter comparison. But when, in 2006, the newspaper headlines screamed that dolphins were "dimwits" and "flippin' idiots," I was shocked. Is this a way to talk about an animal so revered that there are Web domains with "smart dolphin" in their name?

This is not to say that one should believe everything they observe about them. For example, their "smile" is fake (they lack the facial muscles for expressions), and all that science seems to have learned from chatting "dolphinese" with them is that lone male dolphins are keenly interested in female researchers.

Nevertheless, it's going too far to say that dolphins are dumb. Yet this was the claim of Paul Manger, a South African neuroanatomist who said that dolphins' relatively large brains are due to the preponderance of fatty glial cells. Glial cells produce heat, which allows the brain's neurons to do their job in the cold ocean. Manger couldn't resist adding that the intelligence of dolphins and other cetaceans (such as whales and porpoises) is vastly overrated. He offered gems of insight, such as that dolphins are too stupid to jump over a slight barrier (as when they are trapped in a tuna net), whereas other animals will. Even a goldfish will jump out of its bowl, he noted.

If we skip the technicalities—such as that glial cells add connectivity to the brain, and that humans too have many more of these cells than actual neurons—the goldfish remark reminds me of a common strategy to downplay animal intelligence, which is to "demonstrate" remarkable cognitive feats in small-brained species: If a rat or pigeon can do it, or even do it better, it can't be that special. Thus, in order to undermine claims of apes having language skills, pigeons have been trained to have "conversations" by hitting a key that offers information to another pigeon, whereupon the other hits a key marked "Thank you!" These birds have also been conditioned to preen themselves before a mirror, supporting the claim that they are "self-aware."

Clearly, pigeons are trainable. But is this truly comparable to Presley, a dolphin at the New York Aquarium, who, without any rewards or instructions, reacted to being marked with paint by taking off at full speed to a distant part of his tank where a mirror was mounted? There he would spin round and round the way we do in a dressing room, appearing to check himself out.

The mirror test was set up by Diana Reiss and Lori Marino. Their variation on the rouge test was actually more rigorous than such tests on children and apes, because it included a "sham" mark. They first put an invisible mark on two captive-bred dolphins, using water instead of paint, before they used visible marks. For the rouge test it is critical to mark a part of the body (such as right above the eye) that is invisible without guidance of a mirror. The only way the animal should be able to figure out that it has been marked is by seeing itself in the mirror and connecting its reflection with its own body.

The dolphins spent far more time near the mirror, inspecting their reflection, when they had been visibly marked than when they had been sham marked. They seemed to recognize that the mark they saw in the mirror had been put on their own body. Since they hardly paid any attention to marks on other dolphins, it was not as if they were obsessed with marks in general. They were specifically interested in the ones on themselves. Critics complained that the dolphins in this study failed to touch their own body, or rub off the mark, as humans or apes do, but I'm not sure we should hold the absence of self-touching against an animal that lacks the anatomy for it. Until better tests have been designed, it seems safe to let dolphins join the cognitive elite of animals that recognize themselves in a mirror.

Dolphins possess large brains (larger than humans, in fact), and show every sign of high intelligence. Each individual produces its own unique whistle sound by which the others recognize him or her, and there are even indications that they use these sounds to call each other "by name," so to speak. They enjoy lifelong bonds, and reconcile after fights by means of sexy petting (much like bonobos), while males form power-seeking coalitions. They may encircle a school of herring to drive them together in a compact ball, releasing bubbles to keep them in place, after which they pick their food like fruit from a tree.

In captivity, dolphins have been known to outsmart their keepers. Trained to collect debris from her tank, one dolphin was amassing more and more fish rewards until her charade was exposed. She was hiding large items, such as newspapers and cardboard boxes, deep underwater only to rip small pieces from them, bringing these to her trainer one by one.

There are tons of such observations—glia or no glia. Still, I must admit that the whole fat-brain affair, which rightly upset dolphin experts, provided me with some fresh insights. From now on, if I find my goldfish thrashing on the floor, I'll have to congratulate it before dropping it back into its bowl.

With regard to the co-emergence hypothesis, it is important to note the level of dolphin altruism. Does self-awareness go hand in hand with perspective-taking, and do dolphins show the sort of targeted helping known of humans and apes? One of the oldest reports in the scientific literature concerns an incident on October 30, 1954, off the coast of Florida. During a capture expedition for a public aquarium, a stick of dynamite was set off underwater near a pod of bottlenose dolphins. As soon as one stunned victim surfaced, heavily listing, two other dolphins came to its aid: "One came up from below on each side, and placing the upper lateral part of their heads approximately beneath the pectoral fins of the injured one, they buoyed it to the surface in an apparent effort to allow it to breathe while it remained partially stunned." The two helpers were submerged, which meant that they couldn't breathe during their effort. The entire pod remained nearby (whereas normally they'd take off immediately after an explosion), and waited until their companion had recovered. They then all fled in a hurry, making tremendous leaps. The scientists reporting this incident added: "There is no doubt in our minds that the cooperative assistance displayed for their own species was real and deliberate."

Reports of leviathan care and assistance go back to the ancient Greeks. Whales may interpose themselves between a hunter's boat and an injured companion, or capsize the boat. In fact, their tendency to come to the defense of victims is so predictable that whalers take advantage of it. Once a pod of sperm whales is sighted, the gunner only needs to strike one among them. When other pod members encircle the ship, splashing the water with their flukes, or surround the injured whale in a flowerlike formation known as "the marguerite," the gunner has no trouble picking them off one by one. Such sympathy entrapment would work with few other animals.

There is an abundance of stories of human swim-

Two dolphins were seen supporting a third, which they took between them. They buoyed the stunned victim so that its blowhole was above the surface, whereas their own blowholes were not.

mers saved by dolphins or whales, sometimes protected against sharks, or lifted to the surface the same way these animals support one another. I find help that crosses the species barrier most intriguing, including cases of apes saving birds, or of a seal rescuing a dog. The latter happened in public view in a river in Middlesbrough, England, when an old dog that could barely keep its head above the water was nudged to shore by a seal. According to an eyewitness, "A seal popped up out of nowhere. He came behind it and actually pushed him. This dog would not have survived if it hadn't been for that seal."

Of course, helping tendencies hardly evolved to benefit other species, but once in existence they can be freely employed for such purposes. This also holds for human helping when we bestow aid on sea mammals; for example, when angry activists defend whales against hunters (it's hard to imagine them doing the same on behalf of giant jellyfish) or when we rescue stranded whales. People come out in droves to keep them wet and wrap towels around them, and to push them back into the ocean when the tide rises. This requires enormous effort, so is an act of genuine altruism on our species' part.

In one of the more striking descriptions, a whale seemed to understand the human effort, which would suggest perspective-taking. To take advantage of received help is one thing; to actually be grateful is quite another.

On a cold December Sunday in 2005, a female humpback whale was spotted off the California coast, entangled in the nylon ropes used by crab fishermen. She was about fifty feet long. A rescue team was dispirited by the sheer amount of ropes, about twenty of them, some around the tail, one in the whale's mouth. The ropes were digging into the blubber, leaving cuts. The only way to free the whale was to dive under the surface to cut away the ropes. Divers spent about one hour doing so. It was a herculean job, obviously not without risk given the power of a whale's tail. The most remarkable part came when the whale realized it was free. Instead of leaving the scene, she hung around. The huge animal swam in a large circle, carefully approaching every diver separately. She nuzzled one, then moved on to the next, until she had touched all of them. James Moskito described the experience:

> It felt to me like it was thanking us, knowing that it was free and that we had helped it. It stopped about a foot away from me, pushed me around a little bit and had some fun. It seemed kind of affectionate, like a dog that's happy to see you. I never felt threatened. It was an amazing, unbelievable experience.

We'll never know what the whale was saying or whether it was truly grateful, which would require it to understand human effort. Do whales fit the co-emergence hypothesis? Unfortunately (or perhaps fortunately), some animals are just too large for experiments, even relatively simple ones such as the mirror test. This test poses already enough of a challenge with elephants, which are both smaller than whales and land-dwelling.

We were just lucky to have Happy.

She's Happy

The website for a conference titled "What Makes Us Human?" featured videotaped street interviews with Americans about its chosen theme. Typical answers included "Being human means that we care about others," or "We're the only ones sensitive to each other's feelings." These were laypeople, of course, but often I hear the same from fellow scientists. Michael Gazzaniga, a leading neuroscientist, starts an essay about the human brain as follows:

> I always smile when I hear Garrison Keillor say, "Be well, do good work, and keep in touch." It is such a simple sentiment yet so full of human complexity. Other apes don't have that sentiment. Think about it. Our species does like to wish people well, not harm. No one ever says, "have a bad day" or "do bad work" and

keeping in touch is what the cell phone industry has discovered all of us do, even when there is nothing going on.

True, it is human to express such sentiments verbally, as is the invention of the cellphone, but why assume that the sentiments themselves are new? Do apes truly wish one another a bad day at every turn? This remains the standard line, though, even among scientists who appreciate the long evolutionary history of the human brain, including old areas devoted to affection and attachment. I can go on and on offering counterexamples, but am afraid that they're becoming repetitious. I also don't want to give the impression that all I ever see is nice behavior. There's plenty of one-upmanship, competition, jealousy, and nastiness among animals. Power and hierarchy are such a central part of primate society that conflict is always around the corner. Ironically, the most striking expressions of cooperation occur during fights, when primates defend one another, or in their aftermath, when victims receive solace. This means that for many expressions of kindness, something disagreeable had to happen first.

But having said this, the overwhelming evidence that animals, at least some of the time, "wish each other well" is the proverbial elephant in the room during any debate about human nature. I love this English expression, which refers to an obvious truth of massive proportions that is ignored because of its inconvenience. People willfully suppress knowledge most have had since childhood, which is that animals do have feelings and do care about others. How and why half the world drops this conviction once they grow beards or breasts will always baffle me, but the result is the common fallacy that we are unique in this regard. Human we are, and humane as well, but the idea that the latter may be older than the former, that our kindness is part of a much larger picture, still has to catch on.

I am not even particularly interested in demonstrating animal empathy, because for me the critical issue is no longer *whether* they have it, but *how* it works. My suspicion is that it works exactly the same way in humans and other animals, even though humans may add a few complexities. It is the core mechanism that matters, and the circumstances that turn empathy on or off. I am irresistibly drawn, therefore, toward the great beast in the room, wanting to poke and prod it so as to determine what it is made of. Hopefully, not like the six blind men from Indostan, who couldn't agree on any of its parts, but more like a scientist who recruits the knowledge of his day to come up with an account of how one member of a species gets to care about another.

Elephants are well known for this. They don't need a genetic relationship to help one another, such as the aforementioned blind elephant and her friend, both of whom had come to the park from different sources. The same is true in the wild, where unrelated elephants sometimes help one another to their feet as in the description below of a dying matriarch, named Eleanor, on a Kenyan game reserve:

> Eleanor was found with a swollen trunk which she was dragging on the ground. She stood still for a while, then took a few slow small steps before falling heavily to the ground. Two minutes later, Grace [matriarch of a different group], rapidly approached with tail raised and streaming with temporal gland secretion. She lifted Eleanor with her tusks back on to her feet. Eleanor stood for a short while, but was very shaky. Grace tried to get Eleanor to walk by pushing her, but Eleanor fell again facing the opposite direction to her first fall. Grace appeared very stressed, vocalizing, and continuing to nudge and push Eleanor with her tusks.

What fascinates me in these and other cases is how elephants manifest the two-lane path to targeted helping. First of all, there is the arousal, marked by stress signals, such as loud vocalizing, urination, streaming glands, raised tails, and spread ears, which indicates emotional contagion. Second, there is the insightful part, where appropriate assistance is being offered, such as lifting a three-thousand-kilogram fallen comrade to her feet. In a separate case, American wildlife biologist Cynthia Moss witnessed the response after a poacher's bullet had entered the lungs of a young female, Tina. When her knees started to buckle, members of Tina's family leaned into her so as to keep her upright. She died nonetheless, upon which one of the others "went off and collected a trunkfull of grass and tried to stuff it into her mouth."

This last little detail is telling, since it suggests an attempted solution. It may not be the right solution, but isn't it the thought that counts? Elephants normally don't stuff food into one another's mouths, so why start with one who has just died? And why not put it in Tina's ear or, for that matter, her behind? It's the correspondence problem again: The helper seemed aware which part of Tina's body normally would accept food. There are similar observations, such as an older bull bringing water from a nearby spring to a dying companion, spraying it over the other bull's head and ears, and trying to get him to drink. This is highly unusual behavior and suggests an insightful approach to the other's problem.

Thousands of people watched a television nature program in which an adopted baby elephant had slid into a mud hole and couldn't get out. The surrounding elephants became highly agitated. The noise of trumpeting and rumbling was overwhelming, with everyone going into high gear. The matriarch and another female started working on the problem, one of them climbing into the hole on her knees, while the mud was creating deadly suction on the calf. Both females worked together, placing their trunks and tusks underneath the calf until the suction was broken and the calf scrambled out of the hole. When this film clip is shown to a human audience, they clap as soon as the calf stands on dry land, shaking off the mud like a big floppy dog.

Most such observations concern African elephants, which are actually quite different from Asians—they're not only a different species, but a separate genus. Asian elephants do the same, however. Here's one of Josh's e-mails from Thailand:

> I saw an incredible act of targeted helping. An older female, perhaps close to 65, fell down in the middle of the night. It was a very

rainy, muddy jungle environment, difficult for us to walk around, I can only imagine how difficult it was for a tired old female to get up. For hours, mahouts and volunteers alike tried to lift her. In the meantime, her close companion, Mae Mai, an unrelated female of about 45, refused to leave her side. I say refused because mahouts were trying to get her out of the way (tempting her with food). She may have sensed that they were trying to help, because after re-peated tries to lift the fallen female with human hands and with another elephant tethered to her, Mae Mai, in a rather agitated state, got alongside the old female, and with her head, tried to push her up. She repeatedly tried to do so, ending each failed at-tempt with frustrated trunk smacks to the ground and rumbling. She seemed highly committed to staying with her friend.

When the old female died, a few days later, Mae Mai urinated uncontrollably, and started bellowing loudly. When the Mahouts tried to take down a large wooden frame to try and raise the old fe-male, Mae Mai got in the way and wouldn't let the wood anywhere near her dead friend. Mae Mai then spent the next two days wan-dering around the park bellowing at the top of her voice every few minutes, causing the rest of the herd to respond with similar sounds.

Unlike helping among primates, which has been studied from many angles, for elephants we only have stories. But then, those sto-ries come from so many different sources and are internally so consis-tent that I have no doubt that being thick-skinned doesn't keep these animals from being extraordinarily sensitive. In fact, Josh's project in Thailand aims not only to measure how semi–free ranging elephants rally around distressed parties, such as a youngster freaked out by a snake, but also the arousal of the surrounding group. The mournful bellowing by Mae Mai that set off vocalizations by others is a case in point. This gets at emotional contagion, which may be more visible in elephants than most animals, such as when elephants around a frightened herd member stretch their tails and flap their ears. In extreme cases, they empty their bladders and bowels: an outward sign of emotional engagement that is hard to miss.

This also explains our interest in the reaction of elephants to mirrors. We teamed up with Diana Reiss, who had tested the dolphins before, to see if we could get the same thing going with elephants. It sounded simple, until we reflected on the kind of mirror needed. We were definitely thinking big, bigger than an earlier study according to which elephants fail the rouge test. Looking at that study's description, it's not hard to see a few problems. First of all, the mirror was much smaller than an elephant's body. Second, it was placed on the ground at a distance, so that even with the best of vision (which elephants may not have) attention must have been drawn mostly to the reflection of the animal's feet. And finally, the mirror was put outside the enclosure, separated by bars, so that there was no way for the elephant to smell or touch it, or feel behind it, which many animals like to do before interacting further. In short, the setup kept the animal from fully exploring this unusual contraption.

We received excellent cooperation from the Bronx Zoo, which built us what we like to call a "jumbo-sized" mirror. It was a giant plastic mirror of eight by eight feet glued to a metal frame with a sturdy cover, so that we could block its view on days we weren't using it. We didn't want the elephants seeing the mirror unless we could videotape their reactions. The mirror had a tiny lipstick camera in the middle so that we could film everything close up. Most of all, the mirror was elephantproof. The animals could smell and touch it as much as they wanted, and even look behind it, although we felt they were a little too enthusiastic doing so.

Maxine walked up to the mirror and slung her trunk over it, after which she began climbing up, standing on her hind legs so that she could peek over the wall on which the mirror was mounted. Elephants don't climb, as everyone knows, and this was the first time keepers with decades of experience had seen anything like it. The wall withstood the couple of tons leaning on top of it; otherwise our experiment might have ended then and there with a pursuit of Maxine through New York traffic!

After her climbing effort, Maxine adopted a most ridiculous posture, getting completely down on her "elbows" with her large behind and back legs swinging up in the air as she tried to literally stick her entire trunk underneath the mirror. This just goes to show her extreme desire to understand the mirror. On the other hand, at no moment did the elephants treat their reflection as if it were another member of their species. This is remarkable, because even apes and children do so at first sight. Is it possible that smell plays a greater role for elephants, so that it makes no sense for them seeing "another" without accompanying odor cues?

Like the apes, they used the mirror to inspect parts of their bodies that they normally never see. They opened their mouths wide in front of the mirror, feeling into them with their trunks. One elephant pushed her ear forward with her trunk while facing the mirror. They also made strange swinging motions, or walked repeatedly in and out of view of the mirror, as if to make sure that their reflection behaved the same way as they did themselves. This is known as "self-contingency testing," which is typical of apes as well. It was what we were waiting for, as it suggested that the animals had an inkling of what they saw in the mirror.

We prepared for the rouge test, following the same sham-mark procedure as Diana had applied to dolphins. A paint company had provided us with white face paint and a container with exactly the same paint in which a single, odorless component had been changed so that it showed no visible pigmentation. Large X's were painted on both sides of the elephant's head above each eye, on the right with the visible paint, on the left with the invisible one.

Happy, a thirty-four-year-old Asian elephant, did all the right things to indicate that she connected the mirror image with herself. She first walked straight to the mirror, where she spent ten seconds, then moved away. We were disappointed. But without having touched

With a big white X painted on her forehead, Happy walked up to the mirror. She could not see the mark without the mirror, but began feeling and touching it.

the mark, she returned seven minutes later. She moved in and out of the mirror's view a couple of times, until she moved away again. While turning away, she began to feel the visible mark. She then returned to the mirror and, while standing directly in front of it, touched and investigated the mark multiple times with her trunk.

According to our videotapes, Happy directed a dozen touches at the visible mark and none at the sham mark.

The great thing, compared with dolphins, is that the elephant is an animal that can touch itself. By any standard used for apes or children, Happy passed the rouge test. We tested two more elephants, including Maxine, but they failed. This is less surprising than it may seem, because for even the most intensely tested primate, the chimpanzee, the proportion of individuals passing the rouge test is far from 100 percent, and in some studies it is less than half.

To see Happy rhythmically swing her trunk in the direction of the big white cross that she couldn't know about without the mirror, closer and closer, until she began to carefully and precisely touch it, was a sight to behold. We were elated. It was the first indication that elephants have the same capacity for mirror self-recognition that humans, dolphins, and apes have.

For the news media, our scientific report on this discovery appeared at a propitious moment, right after the midterm debacle of the Republican Party in 2006. Their proud symbol, of course, is the elephant. Newspapers couldn't resist cartoons of an injured and bandaged pachyderm sitting in front of a mirror, staring dejectedly at itself. But the funniest opening line came from a widely carried

Associated Press piece: "If you're Happy and you know it, pat your head."

So, it seems that elephants too fit the co-emergence hypothesis. Obviously, we need to better understand their exact level of empathy, and it is paramount that more elephants be subjected to the rouge test. But for the moment I take our evidence as encouraging. Moreover, there is brain research to match, because as it turns out, all mammals with mirror self-recognition possess a rare type of brain cell.

A decade ago, a team of neuroscientists showed that so-called *Von Economo neurons*, or VEN cells, are limited to the hominoid (human and ape) brain. VEN cells differ from regular neurons in that they are long and spindle-like. They reach further and deeper into the brain, making them ideal to connect distant layers. John Allman, a member of the team, thinks that VEN cells are adapted for large brains, adding much-needed connectivity. In the dissection of the brains of many species, these cells were found only in humans and their immediate relatives, but were absent in all other primates, such as monkeys. The cells are particularly large and abundant in our own species, and are found in a part of the brain critical for traits that we consider "humane." Damage to this particular part results in a special kind of dementia marked by the loss of perspective-taking, empathy, embarrassment, humor, and future-orientation. Most important, these patients also lack self-awareness.

In other words, when humans lose their VEN cells, they lose about every capacity that's part of the co-emergence hypothesis. It's unclear if these particular cells themselves are responsible, but it is thought that they underpin the required brain circuitry. Now, if VEN cells play such a vital role in what sets humans and apes apart from the rest of the animal kingdom, the obvious next question is whether they are an absolute requirement. Could other animals, such as dolphins and elephants, possess the same capacities *without* VEN cells?

But we don't need to worry about this, because the latest discovery by Allman's team is that VEN cells are not limited to humans and apes. These neurons have made their independent appearance in only two other branches of the mammalian tree, which happen to be the cetaceans (dolphins and whales) and elephants.

In Its Own Little Bubble

The co-emergence hypothesis offers a nice, tidy story tying together ontogeny, phylogeny, and neurobiology. It's not a story that sets humans apart, even though we have more of everything: more empathy, more VEN cells, and more self-awareness. We go beyond other animals, for example, in that we are self-conscious about our looks, and have an actual opinion about it: Some hate how they look, and some love it. We shave, comb, or decorate ourselves in front of the mirror every day. We not only recognize ourselves, but also care about our appearance. This may not be totally unique (one orangutan at a German zoo had a habit of piling lettuce leaves onto her head before checking out the results in a mirror), but our species is definitely the planet's greatest narcissist.

We are part of a small brainy elite that operates on a higher mental plane than the vast majority of animals. Members of this elite have a superior grasp of their place in the world and a more accurate appreciation of the lives of those around them. But however tidy the story may seem, I'm inherently skeptical of sharp dividing lines. For the same reason that I don't believe in a mental gap between humans and apes, I can't believe that, say, monkeys or dogs have none, absolutely none, of the capacities that we've been discussing. It's just inconceivable that perspective-taking and self-awareness evolved in a single jump in a few species without any stepping stones in other animals.

But let's first look at the differences. In the early 1990s, my coworker Filippo Aureli and I decided to study consolation in monkeys to see if they, like the apes, reassure distressed parties. Both of us had watched hundreds of aftermaths to aggressive conflict in a variety of species, and the setup of those studies had been similar to what we were now planning. The approach is to wait until a spontaneous fight breaks out in a primate group, and then document the events that follow. This method offers unambiguous evidence for consolation in apes, so should do the same in monkeys, provided they have it. At the time, we had no reason to think they wouldn't.

But to our surprise, we found nothing! Whereas reconciliation, in which former opponents come together, occurs in all monkeys studied, consolation is totally absent. How could this be? In fact, the monkey observations were shocking, because we'd see a defeated monkey crouching in a corner, and not even its own family seemed the least bit worried. After more failed monkey studies, Italian scientist Gabriele Schino reasoned that if there's any situation in which one would definitely expect consolation it would be between a mother and her youngest offspring, because this is the closest bond. When Schino tested this on macaques on a large rock at the Rome Zoo, however, his findings were positively baffling. Mothers barely paid attention to their offspring after they had been attacked and bitten, and certainly didn't actively comfort them. This is all the more surprising since macaque mothers do defend their young against aggression, hence recognizing this as an aversive event. And juveniles do run to their mother after an attack, often huddling against her with a nipple in their mouth, seeking comforting contact. They just shouldn't expect their mother to go out of her way to provide it.

There are other indicators of a lack of empathy in monkeys, such as the "exasperating" stories of baboon watchers in the Okavango Delta of Botswana about adults ignoring the fear of youngsters facing a water crossing. Standing panicky at the water's edge, young baboons risk getting killed by predators, yet their mothers rarely return to retrieve them. They just keep traveling. It's not that a baboon mother is entirely indifferent:

> She appears genuinely concerned by its agitated screams. But she seems to fail to understand the cause of this agitation. She behaves as if she assumes that if she can make the water crossing, everyone can make the water crossing. Other perspectives cannot be entertained.

Another Dickensian observation was made during an exceptional flood, which forced the baboons to swim from island to island. One day the adults crossed to another island, leaving almost all of the young stranded behind. The latter were highly stressed, tightly bunched together in a tree, emitting agitated barks. Fieldworkers following the adult baboons could hear the contact calls of the young in the distance, and the adults themselves occasionally oriented toward them but gave no answering barks. The juveniles later managed to swim across and reunite with the troop.

All of this suggests intact emotional contagion but an inability to adopt another's point of view. This is a familiar deficit in many animals as well as young children. Sometimes it takes amusing forms. At home, we have a black-and-white tomcat, Loeke, who is scared to death of strange people, especially men with big shoes. He must have suffered trauma before we adopted him. As soon as visitors enter our house, he races upstairs in total panic and wriggles his way under our bedcovers. He can stay there for many hours at a time, but remains of course extremely visible. We see a bed with a conspicuous bulge and know exactly where Loeke is, but I bet he thinks that since he can't see anybody, nobody can see him. The bulge even purrs if we whisper to it. As soon as we close the front door behind the departing visitors, we look at our watch to see how long it takes Loeke to return to the living. It rarely takes him more than twenty seconds.

But lack of perspective-taking can also take heartbreaking forms, as in the baboons above. I remember visiting a Japanese monkey park where the ranger told me how they needed to keep first-time mothers from entering the hot water springs, since they are likely to drown their babies. Young mothers apparently don't pay sufficient attention to the situation of a baby clinging to their bellies, perhaps thinking that if their own heads are above the surface no one can possibly be in trouble. I have also seen captive monkeys performing dangerous acrobatics in a large spinning wheel in which infants had been playing, forcing the latter to cling to its frame for dear life. An injured young female macaque followed her mother around with a broken, lifelessly dangling arm, without the mother ever adjusting even a tiny bit to her daughter's handicap.

How different from a chimpanzee mother I knew, who accommodated every wish of her juvenile son who had a broken wrist, even to the point of letting him nurse, though he had been weaned years earlier. Until his arm had healed, she put him ahead of his younger sibling. Sensitivity to another's injuries, other than obvious open wounds, requires appreciating how someone's locomotion is hampered. Apes definitely notice this, as do dolphins and elephants. Examples of elephants helping humans are hard to come by, but Joyce Poole offers the account of a matriarch who had attacked a camel herder, breaking the man's leg. The same elephant returned later, and with her trunk and front legs moved him to a shady spot under a tree, where she stood protectively over him. She occasionally touched him with her trunk and chased off a herd of buffaloes. She watched over him an entire day and night, and when a search party showed up the next day she was resistant to let them retrieve their colleague.

Monkeys, in contrast, ignore handicaps and seem to have little clue about grief or loss. A typical study is one by Anne Engh, who measured stress levels of baboons who had witnessed a family member being dragged away by a leopard, lion, or hyena. Predation accounts for a high percentage of deaths, and so Engh had many cases to study. Sometimes her baboons literally heard the fearsome predators crunch the bones of their kin. As one might expect, the ones left behind had an elevated stress level, which Engh measured by extracting corticosterone (a stress hormone) from droppings. She also found that bereaved baboons groomed others more, probably as a way to reduce stress and build new relationships to replace lost ones. About one female, she notes: "So great was her need for social bonding that Sylvia began grooming with a female of a much lower status, behavior that would otherwise be beneath her."

Engh concludes that, like humans, baboons rely on friendly relationships to help them cope with stress. The similarities are indeed obvious, but isn't there also a glaring contrast? Other baboons never changed their behavior toward those who had just lost friends and family, which is a fundamental difference with our own species. We are plenty aware of another's loss, keeping it in mind for years. Chimpanzees too seem sensitive, as when the adolescent daughter of one of our females had been sent off to another facility. We were struck by the enormous amount of grooming others directed at the mother in the ensuing weeks. Chimpanzees offer true solace in a way that we humans understand, whereas bereaved baboons are left to regulate their own internal state. They do seem to have the same needs, but can't expect much consideration from others. No wonder a pioneering baboon watcher characterized this monkey's life as "one continual nightmare of anxiety."

The limited sensitivity of monkeys to others seems due more to cognitive than emotional factors. Monkeys do feel the distress of others but have no good grasp of what's going on with them. They can't step back from the situation to figure out the other's needs. Every monkey lives in its own little bubble.

Yellow Snow

Progress in science has often come from attention to exceptions. The co-emergence hypothesis, too, must have a few.

Monkeys do sometimes show glimpses of the understanding that led to the advanced empathy typical of our own branch of the family tree. These incidents are rare, which is why they are exceptions, but show borderline understanding. The owner of a tame capuchin monkey, for example, told me how his pet once bit a visitor when the latter was trying to hand-feed it grapes. The monkey had given only a tiny nip—no blood was visible—but the woman looked hurt and dropped the grapes. The monkey promptly and gently hugged her with both arms around her neck. It reacted to the woman's distress while ignoring the grapes, which had fallen to the floor. To all present, it looked exactly like human solace. In our own capuchin colony, we have also seen such incidents, but we use such strict criteria to determine if primates have consolation that our capuchins have never satisfied them.

Further incidents in our capuchins concerned females so heavily pregnant that they refused to descend to the ground. Capuchins feel safer higher up. Trays with fruits and vegetables for the colony are put on the floor every evening, however. We have seen close friends and family grab mouth- and handfuls of food (sometimes wrapping food into their tails) and climb up to the gravid female's platform, where they spread it all out in front of them, after which they happily eat together.

Another example has stuck in my mind for decades, ever since I saw a photograph by Hans Kummer, a much-admired Swiss expert of hamadryas baboons. The photo shows a juvenile monkey using an adult's back to climb down from some rocks, accompanied by the following caption by Kummer: "After vainly trying to descend over a difficult passage, a one-year-old has started to scream. His mother finally returns and offers him her back as an additional step." My question here is if such assistance wouldn't have required the mother to appreciate the juvenile's need.

Anindya Sinha, an Indian primatologist, witnessed similar incidents in wild bonnet monkeys. On three separate occasions, a juvenile tried to climb or jump up a parapet but was unsuccessful. After repeated attempts, an adult male who had been watching reached down, grasped the arm of the juvenile, and pulled him up. Only in one case was the male's attention drawn by the juvenile's calls—in the other two cases, the males were unmoved. The helpers were always

Monkeys occasionally assist one another's climbing efforts, as a baboon mother does here with her infant.

alpha males of their troop, but a different individual in every observed instance.

Barbara Smuts relates how adult male baboons sometimes reassure distressed infants by softly grunting at them. By itself, this is already interesting, as it suggests vocal consolation. But Smuts also saw a male do so while, like the monkeys above, he seemed to identify with an infant's intentions. Enjoying the grooming attention of her mother, the male, Achilles, watched the antics of a female infant who tried to climb a sandy mound. When she almost reached the top, she slipped and slid to the bottom. As soon as this happened, Achilles directed some reassurance grunts at her.

Baboons may even express vocal relief when an awkward situation comes to an end, indicating appreciation of the situation others find themselves in. In his usual entertaining style, Robert Sapolsky tells the story of an infant born to a particularly maladroit mother, whose offspring was often forced to cling to her tail:

> One day, as she leapt from one branch to another in a tree with the kid in that precarious position, he lost his grip and dropped ten feet to the ground. We various primates observing proved our close kinship, proved how we probably utilized the exact same number of synapses in our brains in watching and responding to this event, by doing exactly the same thing in unison. Five female baboons in the tree and this one human all gasped as one. And then fell silent, eyes trained on the kid. A moment passed, he righted himself, looked up in the tree at his mother, and then scampered off after some nearby friends. And as a chorus, we all started clucking to each other in relief.

Such observations suggest that baboons do not just react to external signs of distress, such as calls or facial expressions, but that it matters to them that a fallen infant gets up again or that a mother and infant are reunited.

Then there's the curious case of Ahla, a baboon employed by a goat farmer. Ahla knew every mother-lamb relationship in the herd. When mothers and kids were kept in separate barns, she'd come into action as soon as a kid started bleating. She'd go pick it up and carry it under her arm to the other barn. There she would shove the kid underneath the right female for nursing, never making a mistake. This obviously required knowledge of relationships, but perhaps more. I doubt very much that one could train an animal to perform such a task if it lacked any understanding of why baby goats bleat, and indeed Ahla was said not only to be eager but a "maniac" at putting mothers and offspring together.

We should be careful not to overinterpret, of course, but in all of these cases monkeys give indications of perspective-taking. The fact that they do not do so consistently, and perhaps only in relation to a narrow set of circumstances, may explain why systematic studies of targeted helping or consolation in monkeys typically come up empty-handed. These primates just act like this too sporadically.

The only natural situation that may be common enough for a serious study is so-called bridging behavior. Some South American primates have prehensile tails and create live bridges between trees for their older offspring (younger offspring simply cling to their mother during travel). When moving through the canopy, the mother holds on with her tail to one tree while grasping a branch of another with her hands, hanging suspended like this until her offspring has crossed over. Short of moving over the forest floor, which is too dangerous, the young would be unable to travel without these maternal bridges. It is an intriguing everyday helping act that begs careful observation since it may reveal the degree to which monkeys take another's abilities into account. When I followed capuchins in the forest, I didn't see any of it, but this is because these monkeys are such great jumpers. The larger and heavier primates, such as howler and spider monkeys, do produce bridges, sometimes even for unrelated juveniles. Females are said to adjust their behavior to new situations, such as an offspring with a broken arm or a forest with unusually wide gaps between its trees.

Juveniles sometimes cry to solicit their mother's help, but more often than not she bridges spontaneously. She decides which distances her offspring can or cannot handle, and that obviously changes with age. What a wonderful setup to test perspective-taking, especially since the monkeys' behavior can be directly compared with that of apes, such as orangutans, who do the same (without tail) by swinging small trees until they can pull them together for their young to cross over. Do orangs do so with greater insight into their offspring's needs than South American monkeys have? I'd expect so, but someone will need to make sure.

For the moment, the proposed empathy differences between monkeys and apes remain intact, although they seem less absolute than suggested. So, what about the other part of the co-emergence equation: mirror responses? Here we do seem to have a radical difference. Despite many, many attempts to test monkeys with mirrors under all sort of circumstances, they just don't pass the rouge test. Positive results have been claimed, but none have stood up to scrutiny. It's widely accepted, therefore, that monkeys don't pass. This doesn't mean that they find mirrors totally baffling or that they lack any sense of self. They must have some self-awareness, because no animal can do without it. Every animal needs to set its body apart from the surrounding environment and have a sense of agency. You wouldn't want to be a monkey up in a tree without awareness of how your own body will impact a lower branch on which you intend to land. Or if you're engaged in rough-and-tumble fun with another, what would be the point of gnawing on your own foot? Monkeys never make this mistake, happily gnawing on their partner's foot instead. The self is part of every action an animal—any animal—undertakes.

In a study with the intriguing subtitle "Tales of Displaced Yellow Snow," Marc Bekoff investigated the reactions of his dog, Jethro, to patches of discolored snow outside Boulder, Colorado. With gloved hands, Bekoff would pick up a patch of urine-soaked snow freshly marked by another dog, out of view of Jethro, and bring it to a place next to a bicycle lane, where Jethro would discover it. Bekoff undertook this experiment—which must have looked positively weird to onlookers—to see if Jethro could tell his own markings from those of other dogs. Of course, he could. He was far less inclined to mark over his own urine than that of other dogs. Self-recognition takes many forms.

When it comes to mirrors, too, things are less clear-cut than they seem. Monkeys, for example, are able to use a mirror to locate food. If you hide food that can be found only by using a mirror to look around a corner, a monkey will have no trouble reaching for it. Many a dog can do the same: Holding up a biscuit behind them while they watch you in a mirror makes them promptly turn around. But even though dogs understand mirror basics, try to mark them surreptitiously, or try to do so with a monkey, and all of a sudden they are at a loss. It is specifically the relation with their own body, their own self, that they fail to grasp.

On the other hand, the standard claim that monkeys see a stranger in the mirror is questionable. To test it, we conducted a simple experiment that, surprisingly, no one had ever tried before. We compared how monkeys react to their mirror image with how they react to strangers, testing capuchins in front of a Plexiglas panel behind which they faced either a familiar monkey, a stranger of their own species, or a mirror. Could they tell the difference?

They treated their mirror image quite differently from real monkeys, and did so within seconds. They didn't need any time to notice the difference. Apparently, there are many levels of mirror understanding, and our monkeys never confused their reflection with another monkey's. For example, they reacted to strangers by turning their backs, barely glancing at them, whereas with their own reflections they made prolonged eye contact as if they were thrilled to see themselves. Some of the tested monkeys had young offspring, and since we never separate mothers and infants, the little ones were present during tests. For me, the most telling finding of the whole study was that when there was a stranger on the other side, mothers held their infants tight, not letting them wander around. During mirror tests, on the other hand, they let their kids roam freely. Given how conservative mother monkeys are when it comes to danger, this convinced me more than anything that their reflection was no stranger to them.

Both with regard to empathy and self-recognition the lines separating species remain intact, yet they are perhaps a bit vaguer than at first appears. It's always the same story: We start out postulating sharp boundaries, such as between humans and apes, or between apes and monkeys, but are in fact dealing with sand castles that lose much of their structure when the sea of knowledge washes over them. They turn into hills, leveled ever more, until we are back to where evolutionary theory always leads us: a gently sloping beach. I do believe that the co-emergence hypothesis offers useful clues about the steepness of the beach's slope, but wouldn't be surprised if this turns out to be a temporary obsession. We're already facing an avalanche of new studies, not only on monkeys, but also large-brained birds and canids, that address perspective-taking, consolation, and mirror self-recognition. The leveling waves are in full swing.

Take the magpie. Applying the same sham-mark design as used with dolphins and elephants, a recent study has shown mirror self-recognition in magpies. Now, mind you, the magpie isn't just any bird: It is a corvid, a family that includes crows, ravens, and jays, endowed with exceptionally large brains. Put in front of a mirror, magpies will try to remove a tiny colored sticker attached to their throat feathers. They will keep scratching with their foot until the mark is gone, but will leave a black mark alone probably because it doesn't stand out against their black throat. They also won't do any frantic scratching if there's no mirror to see themselves in. The videos of Helmut Prior and his fellow German scientists are quite compelling. When I saw them in an audience full of corvid experts, I could sense great pride in "their" birds.

The finding is interesting in relation to the magpie's reputation. As a child, I learned never to leave small shiny objects, such as teaspoons, unattended outdoors since these raucous birds will steal anything they can put their filthy beaks on. This folklore even in-spired a Rossini opera, *La Gazza Ladra* ("The Thieving Magpie"). Nowa-days, this view has been replaced with one more sensitive to ecological balance, in which magpies are depicted as murderous robbers of the nests of innocent songbirds. Either way, they are despised as black-and-white gangsters.

But no one has ever called a magpie stupid. For me, the big question is whether its self-recognition supports or undermines the co-emergence hypothesis. For the moment, I believe the for-mer. Perspective-taking may be critically important for a species that plunders the nests of others and steals from humans. The capacity may be even more useful in relation to its own kind. Magpies cache food and undoubtedly steal from one another the way jays also do. This requires watching to see if you're being watched, because if another bird has seen where you hide your food, it's bound to disappear. This topic has been studied in jays ever since a British scientist, Nicky Clayton, observed scrub jays over lunchtime at the University of California at Davis. Clayton noticed the fierce competition for left-over scraps, which the birds would cache away from one another. Some jays went further, however, returning to rebury their treasures once their rivals had left the scene.

Follow-up research with Nathan Emery at Cambridge University led to the intriguing claim that "it takes a thief to know a thief." Jays apparently extrapolate from their own experiences to the intentions of others, so that those who in the past have misappropriated the caches of others are especially keen on keeping the same thing from happening to themselves. Perhaps this process, too, requires the abil-ity to parse the self from the other. As the ultimate bird thief, the magpie may have an even greater need to guess the intentions of oth-ers. Curiously, their self-recognition may therefore relate to a life of crime.

At the very least, these new insights into *la gazza ladra* offer fresh meaning to its love affair with reflective items.

Pointing Primates

Nikkie, a chimpanzee, once showed me how to manipulate attention. He had gotten used to my throwing wild berries across the moat at the zoo where I worked. One day, while I was recording data, I had totally forgotten about the berries, which hung on a row of tall bushes behind me. Nikkie hadn't. He sat down right in front of me, locked his red-brown eyes into mine, and—once he had my attention—abruptly jerked his head and eyes away from mine to fixate with equal intensity on a point over my left shoulder. He then looked back at me and repeated the move. I may be dense compared with a chimpanzee, but the second time I turned around to see what he was looking at, and spotted the berries. Nikkie had indicated what he wanted without a single sound or hand gesture.

That simple act of communication went against an entire body of literature that links pointing to language and has therefore no room for nonlinguistic creatures. Pointing, a so-called deictic gesture, is defined as drawing another's attention to an out-of-reach object by locating the object in space for the other. There's obviously no point to pointing unless you understand that the other has *not* seen what you have seen, which involves realizing that not everyone has the same information. It's yet another example of perspective-taking.

Humans point all the time, and automatically follow the pointer's attention.

Inevitably, academics have surrounded pointing with heavy theoretical artillery. Some have focused on the typically human gesture with outstretched arm and index finger. That gesture has been linked to symbolic communication, which calls up the image of early humans walking around on the savanna, pointing and assigning words to objects: "Let's call the animal over there a zebra, and let's call this here your belly button." Yet doesn't such a scenario imply that our ancestors understood pointing prior to the evolution of language? If so, the idea that our nonlinguistic relatives point shouldn't upset anyone.

The first step is to move away from silly Western definitions, such as the one requiring an outstretched index finger. In our own species, too, a lot of pointing is done without the hands, and in many parts of the world hands are in fact taboo. In 2006, a major health organization advised American business travelers to refrain from finger-pointing altogether, since so many cultures consider it rude. Among Native Americans, for example, approved forms of pointing involve pursed lips, chin movements, a nod with the head, or pointing with knees, feet, or shoulders. I've even heard in-jokes about it, such as the hunting dog whom the white owner brought back to its Indian trainer for retraining, because the dog only knew to indicate game by puckering its lips.

Most people have felt the need to warn companions at a party that the dreaded character X has just entered the room and is heading their way. In such a situation, you don't simply point or shout, even though those are natural inclinations. No, you raise your eyebrows at your companion, jerk your head a bit in the direction of the approaching X, and maybe clamp your lips firmly together to warn your companion to stay mum.

We should take a broad view, therefore, of what constitutes pointing. After all, we have bred dogs (called "pointers") to freeze into a particular stance to indicate a covey of quail. Monkeys, too, often point with their whole bodies and heads when they recruit allies during fights. If monkey A threatens monkey B, B may walk over to his usual protector, C. Sitting next to C, he then looks at him, jerks his head with grunts and threats toward A, and repeats this back-and-forth many times, as if telling C: "Look at that guy—he's bothering me!" Among macaques, an aggressor points with lifted chin and staring eyes at the opponent, in between conspicuously glancing at the ally. Among baboons, the same behavior is so repetitive and exaggerated that fieldworkers call it "head-flagging." The goal is to make absolutely clear to the ally where one's adversary is.

In Emil Menzel's classic studies of knowledge attribution, one chimp knew hidden food or danger, whereas others didn't. The others quickly grasped whether the concealed object was attractive or frightening, and its approximate location, by watching the first chimp's body language. Menzel considered body orientation a highly accurate indicator, especially for an observer in a tree or other high point, adding that "it is primarily a bipedal animal such as man—whose body posture is a much less accurate 'pointer' than the posture of a quadrupedal animal—who actually needs to extend some appendage to indicate direction precisely."

A widely employed criterion for the intentionality of pointing is that the pointer checks the results of his actions. The pointer should make sure, by looking back and forth between the object he is pointing at and his partner, that the partner is paying attention and the pointer is not pointing for nothing. Nikkie did this by locking his eyes with mine. A number of recent experiments have investigated this issue in great apes, using manual pointing—not because this is the most natural way for them to point, but because captive apes readily learn that the gesture is most successful in getting a human response.

At the Yerkes Primate Center, David Leavens worked with chimps who regularly see people walk by. It is logical for the apes to learn how to draw attention to things they want, such as a piece of fruit that has dropped outside of their cage. This was tested systematically by placing food at certain locations. As it turns out, two-thirds of more than one hundred chimps gestured to the experimenter. A few did so by stretching out an open palm. Most, however, used the whole hand to point at a banana outside their cage, although no one had ever trained them to point this way. A few even pointed with an index finger.

There were signs that the apes monitored the effect of their gestures the same way children do. An ape would make eye contact with the human and then point while alternating its gaze between the food and the human. One chimp, afraid to be misunderstood, pointed first with her hand at a banana and then with a finger at her mouth.

Just to illustrate how creative chimps can be, a typical incident happened to me not long ago. A young female, Liza, at the field station, grunted at me from behind the mesh and kept looking at me with shiny eyes (indicating she knew something exciting), alternating with pointed stares into the grass near my feet. I couldn't figure out what she wanted, until she spit into the grass. From the trajectory, I noticed a small green grape. When I gave it to her, Liza ran to another spot and repeated her performance. She proved a very accurate spitter, and altogether got three rewards this way. Liza must have memorized the places where a caretaker had dropped the fruits; she then found me to do her bidding.

Additional evidence comes from what may be the most telling study of referential signaling by apes; that is, signaling with reference to external objects or events, conducted by Charles Menzel (son of Emil) at the Language Research Center of Georgia State University. Menzel let a female chimpanzee, named Panzee, watch him from her enclosure while he hid an object in the surrounding forest. When the caretakers arrived the next morning, they didn't know what Menzel had done. Panzee recruited them by means of pointing, beckoning, panting, and calling, until they had located the item in the forest and given it to her. She was very insistent and explicit in her directions, occasionally using finger-pointing.

It's highly unusual for apes to point things out for one another, though. Perhaps they don't need signals as overt as the ones we employ since they are such incredibly astute readers of body language. But there do exist a few reports of manual pointing, one observed by myself in the 1970s:

> The threatened female challenges her opponent with a high-pitched, indignant bark, at the same time kissing and making a fuss of the male. Sometimes she points at her opponent. This is an unusual gesture. Chimpanzees do not point with a finger but with

their whole hand. The few occasions on which I have seen them actually point have been when the situation was confused; for example, when the third party had been lying asleep or had not been involved in the conflict from the start. On such occasions the aggressor would indicate her opponent by pointing her out.

Another account involves wild bonobos in dense forest, in the Democratic Republic of the Congo, studied in 1989 by Spanish primatologist Jordi Sabater-Pi. One ape alerted his travel companions to the presence of hidden scientists:

> Noises are heard coming from the vegetation. A young male swings from a branch and leaps into a tree. . . . He emits sharp calls, which are answered by other individuals who are not visible. He points—with his right arm stretched out and his hand half closed except for his index and ring fingers—to the position of the two groups of camouflaged observers who are in the undergrowth (30 meters apart). At the same time he screams and turns his head to where the other members of the group are. The same individual repeats the pointing and calling sequence twice. Other neighboring members of the group approach. They look towards the observers. The young male joins them.

In both examples, the context was entirely appropriate (the apes pointed at objects that others had failed to notice), and the behavior was accompanied by visual checking of its effects. Also, the pointing disappeared once the others had looked or walked into the indicated direction. Among our chimpanzees at the field station we have seen similar cases, one of which we captured on video: a finger-pointing female with fully stretched arm, looking almost accusatory as she aimed her gesture at a male who had just slapped her.

One difference with human pointing is that primates show this behavior so rarely, and only in relation to what they consider urgent matters, such as food or danger. Free sharing of information is unusual among them, whereas we do it all the time. Two people may walk through a museum and draw each other's attention to item after item while discussing ancient artifacts; a child will point at a balloon in the air to make sure her parents don't miss it, or someone will point at my bicycle lamp at night to let me know it isn't working. The absence of such behavior in other primates has been taken to indicate a major cognitive gap, but my own guess is that it has more to do with lack of motivation. After all, it's hard to see why a species that points to food to get humans to fetch it couldn't do the same with regard to inedible objects. If they don't do so, it must be because they don't feel like it.

But here too exceptions occur. Bonobos utter small peeps—high-pitched, brief vocalizations—when they discover something of interest. Watching bonobos at the San Diego Zoo day in and day out, I was struck by one group of juveniles, which every morning after their release went around a large grassy enclosure, emitting peeps for lots of things (that I rarely could identify), whereupon the others would hurry over to check out the indicated object. Perhaps they were drawing attention to discoveries such as insects, bird droppings, flowers, and the like. Given how the others reacted, the peeps conveyed something like "Come look at this!"

In our chimpanzee colony, active information-sharing occurred when the always intrepid Katie was digging into the dirt underneath a large tractor tire. Katie uttered soft "hoo" alarm calls during the job, then pulled out something wriggly that looked like a live maggot. She was holding it away from herself between her index and middle finger, a bit like a cigarette. First she sniffed it and then turned and showed it to others, including her mother, holding it up with an outstretched arm. Katie then dropped the object and moved off. Her mother, Georgia, came over and started digging at the same spot. She pulled out something, sniffed at it, and immediately started to alarm bark, but much louder than her daughter had done. She dropped the object and sat down at a distance, still alarm-calling to herself. Then the young daughter of Georgia's sister (Katie's niece) went to the same exact spot under the tire and pulled out something. She walked bipedally over to Georgia and held the object up for her. Georgia now alarm-called even more intensely. Then it was all over.

Vicky Horner, who observed this sequence, thought that the object of interest must have been a disgusting dead rat or something else under the tire that the apes could smell and that was covered with maggots, but she couldn't be sure.

What makes information-sharing interesting is that it relies on the same comparison of one's own perspective with that of someone else—detecting something that others need to know about—which also underlies advanced empathy. Perhaps the capacity to do so appeared only in those few species with a strong sense of self, which is also what permits two-year-old children to engage in such behavior. But very soon children go further, and information-sharing becomes an obsession with them. They feel the need to comment on everything, and ask about everything. This seems uniquely human and may have to do with our linguistic specialization. Language requires consensus, which can't be achieved without continual comparing and testing.

If I point out an animal in the distance and say "zebra," and you disagree, saying "lion," we have a problem that, at other times, may get us into deep trouble. It's a uniquely human problem, but so important to us that deictic gestures and language evolution are closely intertwined.

6

Fair Is Fair

*Every man is presumed to seek what is good for himselfe naturally,
and what is just, only for Peaces sake, and accidentally.*

—THOMAS HOBBES, 1651

In the early spring of 1940, with Nazi troops marching on the city, the population of Paris packed up and fled. In *Suite Française*, eyewitness Irène Némirovsky—who would perish in Auschwitz a few years later—describes how the wealthy lost everything in this mass exodus, including their privileges. They would start out with servants and cars, packing their jewelry and carefully wrapping up their precious porcelain, but soon their servants would abandon them, gas would run out, the cars would fall apart, and the porcelain, well, who cares when survival is at stake?

Even though Némirovsky herself came from a wealthy background, between the lines of her novel the reader detects a certain satisfaction that at times of crisis class differences fade. There's an element of justice to the fact that when everybody suffers, the upper crust suffers, too. High-handed manners that normally would get aristocrats into a hotel, for example, don't help when all rooms are filled to the brim, and an aristocrat's stomach responds the same to being empty as everyone else's. The only difference is that the upper class feels the indignity of the situation more keenly:

> He looked at his beautiful hands, which had never done a day's work, had only ever caressed statues, pieces of antique silver, leather books, or occasionally a piece of Elizabethan furniture. What would he, with his sophistication, his scruples, his nobility—which was the essence of his character—what would he do amid this demented mob?

Passages like these seem almost designed to induce the opposite of empathy: Schadenfreude. They exploit the secret satisfaction we take in the misfortune of the rich. Never the poor, which is telling. We humans are complex characters who easily form social hierarchies, yet in fact have an aversion to them, and who readily sympathize with others, unless we feel envious, threatened, or concerned about our own welfare. We walk on two legs: a social and a selfish one. We tolerate differences in status and income only up to a degree, and begin to root for the underdog as soon as this boundary is overstepped. We have a deeply ingrained sense of fairness, which derives from our long history as egalitarians.

Hunting Hare or Stag?

That we differ from apes in our attitude toward social hierarchy struck me most acutely when chimpanzees failed to react to events that cracked me up.

The first time occurred when a powerful alpha male, Yeroen, was in the midst of an intimidation display with all of his hair on end. This is nothing to make fun of. All other apes watch with trepidation, knowing that a male in this testosterone-filled state is keen to make his point: He is the boss. Anyone who gets in the way risks a serious beating. Moving like a furry steam locomotive that would flatten everybody and everything, Yeroen went with heavy steps up a leaning tree trunk that he often stamped on in a steady rhythm until the whole thing would shake and creak to amplify his message of strength and stamina. Every alpha male comes up with his own special effects. It had rained, however, and the trunk was slippery, which explains why at the peak of this spectacle the mighty leader slipped and fell. He held on to the trunk for a second, then dropped to the grass, where he sat looking around, disoriented. With a "the show must go on" attitude, he then wrapped up his performance by running straight at a group of onlookers, scattering them amid screams of fear.

Even though Yeroen's plunge made me laugh out loud, as far as I could tell none of the chimps saw anything remotely comical. They kept their eyes on him as if this was all part of the same show, even though, clearly, it was not how Yeroen had intended things to go. A similar incident happened in a different colony, when the alpha male picked up a hard plastic ball during his display. He often threw this ball up in the air with as much force as possible—the higher the better—after which it would come down somewhere with a loud thud. This time, however, he threw the ball up and checked around with a puzzled expression because the ball had miraculously disappeared. He didn't know that it was returning to earth by the same trajectory he had launched it on, landing with a smack on his own back. This startled him, and he broke off his display. Again, I found this a rather amusing sight, but none of the chimps showed any reaction that I could tell. Had they been human, they'd have been rolling around, holding their bellies with laughter, or—if fear kept them from doing so—they'd have been pinching one another, turning purple in an attempt to control themselves.

Attributed to St. Bonaventura, a thirteenth-century theologian, the saying "The higher a monkey climbs, the more you see of its behind" illustrates what we think of higher-ups. In fact, the saying applies better to humans than monkeys. My own reactions fit this mold in that I perceived the incongruity of a top individual making a fool of himself during a show of pomp and power, the same way that we can't suppress a laugh when political leaders make embarrassing gaffes or find themselves in their underpants attached to a dancing pole at a strip club. The Australian politician to whom the latter happened during a police raid said it had taught him two lessons: "Don't let anyone handcuff you to a post and make sure you always wear clean underwear."

Our species has a distinctly subversive streak that ensures that, however much we look up to those in power, we're always happy to bring them down a peg. Present-day egalitarians, who range from hunter-gatherers to horticulturalists, show the same tendency. They emphasize sharing and suppress distinctions of wealth and power. The would-be chief who gets it into his head that he can order others around is openly told how amusing he is. People laugh in his face as well as behind his back. Christopher Boehm, an American anthropologist interested in how tribal communities level the hierarchy, has found that leaders who become bullies, are self-aggrandizing, fail to redistribute goods, or deal with outsiders to their own advantage quickly lose respect and support. The weapons against them are ridicule, gossip, and disobedience, but egalitarians are not beyond more drastic measures. A chief who appropriates the livestock of other men or forces their wives into sexual relations risks death.

Social hierarchies may have been out of fashion when our ancestors lived in small-scale societies, but they surely made a comeback with agricultural settlement and the accumulation of wealth. But the tendency to subvert these vertical arrangements never left us. We're born revolutionaries. Even Sigmund Freud recognized this unconscious desire, speculating that human history began when frustrated sons banded together to eliminate their imperious father, who kept all women away from them. The sexual connotations of Freud's origin story may serve as metaphor for all of our political and economic dealings, a connection confirmed by brain research. Wanting to see how humans make financial decisions, economists found that while weighing monetary risks, the same areas in men's brains light up as when they're watching titillating sexual images. In fact, after having seen such images, men throw all caution overboard and gamble more money than they normally would. In the words of one neuroeconomist, "The link between sex and greed goes back hundreds of thousands of years, to men's evolutionary role as provider or resource gatherer to attract women."

This doesn't sound much like the rational profit maximizers that economists make us out to be. Traditional economic models don't consider the human sense of fairness, even though it demonstrably affects economic decisions. They also ignore human emotions in general, even though the brain of *Homo economicus* barely distinguishes sex from money. Advertisers know this all too well, which is why they often pair expensive items, such as cars or watches, with attractive women. But economists prefer to imagine a hypothetical world driven by market forces and rational choice rooted in self-interest. This world does fit some members of the human race, who act purely selfishly and take advantage of others without compunction. In most experiments, however, such people are in the minority. The majority is altruistic, cooperative, sensitive to fairness, and oriented toward community goals. The level of trust and cooperation among them exceeds predictions from economic models.

We obviously have a problem if assumptions are out of whack with actual human behavior. The danger of thinking that we are nothing but calculating opportunists is that it pushes us precisely toward such behavior. It undermines trust in others, thus making us cautious rather than generous. As explained by American economist Robert Frank,

> What we think about ourselves and our possibilities determines what we aspire to become.... The pernicious effects of the self-interest theory have been most disturbing. By encouraging us to expect the worst in others it brings out the worst in us: dreading the role of the chump, we are often loath to heed our nobler instincts.

Frank believes that a purely selfish outlook is, ironically, not in our own best interest. It narrows our view to the point that we're reluctant to engage in the long-term emotional commitments that have served our lineage so well for millions of years. If we truly were the cunning schemers that economists say we are, we'd forever be hunting hare, whereas our prey could be stag.

The latter refers to a dilemma, first posited by Jean-Jacques Rousseau in *A Discourse on Inequality,* that is gaining popularity among game theorists. It's the choice between the small rewards of individualism and the large rewards of collective action. Two hunters need to decide between each going off alone after hare or both sticking together and bringing home stag—much bigger game even if halved. In our societies, we have successfully formalized layers of trust (allowing us to pay with credit cards, for example, not because the shop owner trusts us, but because he trusts the card company, which in turn trusts us), which means that we engage in complex stag hunts. But we don't do so unconditionally. With some people we embark on cooperative ventures more willingly than with others. Productive partnerships require a history of give-and-take and proven loyalty. Only then do we accomplish goals larger than ourselves.

The difference is dramatic. In 1953, eight mountaineers got into trouble on K2, one of the highest and most dangerous peaks in the world. In −40° Celsius temperatures, one member of the team developed a blood clot in his leg. Even though it was life-threatening for the others to descend with an incapacitated comrade, no one considered leaving him behind. The solidarity of this group has gone down in history as legendary. Contrast this with the recent drama, in 2008, in which eleven mountaineers perished on K2 after having abandoned their common cause. One survivor lamented the drive for self-preservation: "Everybody was fighting for himself, and I still don't understand why everybody were [sic] leaving each other."

The first team was hunting stag, the second hare.

Eye-Poking Trust

What is a company retreat nowadays without a trust-boosting game? One man stands on a table with his back toward his colleagues, who are ready to catch him when he drops backward into their arms. Or one woman verbally guides a blindfolded co-worker through an open area with scattered objects that represent a minefield. Trust-building breaks down barriers between individuals, instilling faith in one another, which in turn prepares them for joint enterprises. Everyone learns to help everyone else.

These games are nothing, however, compared to those of capuchin monkeys in Costa Rica. In fact, humans would never be allowed to play the monkeys' games; any lawyer would counsel against them. What these little monkeys do high up in the trees is so absurd that I couldn't believe it until I saw the videos that Susan Perry, an American primatologist, shows nervously empathetic audiences. Two typical games are "hand-sniffing" and "eyeball poking."

In the first, two monkeys sit opposite each other on a branch, both inserting a finger ever deeper into the other's nostril until the finger vanishes up to the first knuckle. Swaying gently, they sit like this with expressions on their faces described as "trancelike." The monkeys are normally hyperactive and sociable, but hand-sniffers sit apart from the group, concentrating on each other for up to half an hour.

Even more curious is the second game, in which one monkey inserts almost a whole finger between the other's eyelid and eyeball. Monkey fingers are tiny, but relative to their eyes and noses they aren't any smaller than ours. Also, their fingers have nails, which obviously aren't particularly clean, so this behavior potentially scratches the cornea or causes infections. Now, the monkeys really need to sit still; otherwise someone may lose an eye. These games are most painful to watch! The pair keeps its posture for minutes, while the one whose eye is being poked may stick a finger into the other's nostril.

What purpose these weird games serve is unclear, but one idea is that the monkeys are testing their bonds. This explanation has also

been offered with respect to human rituals in which we make ourselves vulnerable. Tongue-kissing, for example, carries the risk of disease transmission. Intimate kissing is either pleasurable or totally
disgusting depending on the partner: Engaging in it thus says a lot
about how we perceive the relationship. In couples, kissing is thought
to test the love, enthusiasm, even faithfulness of the partner. Perhaps
capuchin monkeys, too, are trying to find out how much they really like
each other, which may then help them decide who can be trusted to
support them during confrontations within the group. A second explanation is that these games help the monkeys reduce stress, of which
they have no shortage. Their group life is full of drama. During eye-
poking or hand-sniffing, they seem to enter an unusually calm, dreamy
state. Are they exploring the borderline between pain and pleasure,
perhaps releasing endorphins in the process?

I speak of "trust games" because at the very least a high level of
trust is needed before you'd let anyone poke your eye. Exposing yourself to risk on the assumption that others won't take advantage is the
deepest trust there is. What these monkeys seem to be telling each
other—similar to humans when they drop backward into the arms of
others—is that based on what they know about each other, they have
faith that all will end well. This is obviously a wonderful feeling, one
we mostly appreciate with friends and family.

Animals develop such relationships quite readily, also between
species. As pets, they do so with us, so that we can hold them upside
down or stuff them under our sweater—scary moves that they won't
accept from strangers. Or, conversely, we stick an arm into the mouth
of a large dog—a carnivore designed to take a chunk out of it. But animals also learn to trust one another. In an old-fashioned zoo, a monkey kept in the same enclosure as a hippopotamus acted as dental
cleaner. After the hippo had eaten its fill of cucumbers and heads of
salad, the little monkey would approach and tap the hippo's mouth,
which would open wide. It was obvious that they had done this before. Like a mechanic under the hood of a car, the monkey would lean
in and systematically pull food remains from between the hippo's

teeth, consuming whatever he pulled out. The hippo seemed to enjoy the service, because he'd keep his mouth open as long as the monkey was busy.

The risk the monkey took wasn't as great as it might seem. A hippo may have a huge mouth with dangerous teeth, but it's hardly a carnivore. It's much trickier to perform such a job on an actual predator, but this too happens. Cleaner wrasses are small marine fish that feed on the ectoparasites of much larger fish. Each cleaner owns a "station" on a reef with a clientele, which come to spread their pectoral fins and adopt postures that offer the cleaner a chance to perform its trade. Sometimes cleaners are so busy that clients wait in line. It's a perfect mutualism. The cleaner nibbles the parasites off the client's body surface, gills, and even the inside of its mouth.

In a sign of mutual trust, a monkey at a zoo cleans a hippo's mouth.

The cleaner fish trusts that it will be allowed to nibble parasites and that the big fish won't cut his career short. But the big fish needs trust, too, because not all cleaners do an honest job. They sometimes take a quick bite out of the client, feeding on healthy skin. This causes the large fish to jolt or swim away. According to Swiss biologist Redouan Bshary, who has followed these interactions in the Red Sea, cleaners hurry to repair damaged relationships and lure back their clients. They offer a "tactile massage" by moving around the big fish, tickling its belly with their dorsal fins. The other likes this so much that he becomes paralyzed: He starts drifting around motionlessly, bumping into the reef. The massage seems to restore trust, because the big fish usually stays around for further cleaning.

The only fish that cleaners never cheat are the large predators. With them, they wisely adopt what Bshary calls an "unconditionally cooperative strategy." How do those little fish know which clients might eat them? They are unlikely to know this from firsthand experience, which by definition is terminal. Is it because they have seen those fish eat others, or—like Little Red Riding Hood—have they noticed what big teeth they have? We usually don't assume much knowledge in fish, but this is only because we underestimate them, as we do most animals.

Trust is defined as reliance on the other's truthfulness or cooperation, or at least the expectation that the other won't dupe you. This seems a perfectly fine characterization of how cleaner fish must relate to their hosts when they enter their gills or mouth, or the basis on which capuchin monkeys decide with whom to play their eye-poking games. Trust is the lubricant that makes a society run smoothly. If we had to test everyone all the time before doing something together, we'd never achieve anything. We use past experiences to decide whom to trust, and sometimes rely on generalized experience with members of our society.

In one experiment, two persons each received a small amount of money. If one gave up its amount, the other's money would be doubled. The partner was in the same situation. So the best would be for both to give up their money, because then both would gain. These people didn't know each other, however, and weren't allowed to talk. Moreover, the game was played only once. Under these circumstances, it seems smarter to just keep what you have, because you can't count on the other. Yet some people gave up their money anyway, and if both members of a pair did so they obviously had a better income than the rest. The main message of this study, and many others, is that our species is more trusting than predicted by rational-choice theory.

Confidence in others may be fine in a one-shot game with little money, but in the long run we need to be more careful. The problem with any cooperative system is that there are those who try to get more out of it than they put in. The whole system will collapse if we don't put a halt to freeloading, which is why humans are naturally cautious when they deal with others.

Strange things happen if this caution is lacking. A tiny proportion of humans is born with a genetic defect that makes them open and trusting to anyone. These are patients with Williams syndrome, a condition caused by the nonexpression of a relatively small number of genes on the seventh chromosome. Williams syndrome patients are infectiously friendly, highly gregarious, and incredibly verbose. Ask a teen with autism or Down syndrome "What if you were a bird?" and you may not get much of an answer, but the Williams child will say, "Good question! I'd fly through the air being free. If I saw a boy I'd land on his head and chirp."

Even though it is hard to resist these charming children, they lack friends. The reason is that they trust everyone indiscriminately and love the whole world equally. We withdraw from such people since we don't know whether we can count on them. Will they be grateful for received favors, will they support us if we get into a fight, will they help us achieve our goals? Probably none of the above, which means that they don't have anything that we're looking for in a friend. They also lack the basic social skill of detecting the intentions of others: They never assume wrong intentions.

Williams syndrome is an unfortunate experiment of nature that shows that just being friendly and trusting, which is what these patients excel at, is not sufficient for lasting ties: We expect people to be discriminating. That a small number of genes can cause such a deficit tells us that the normal tendency to be circumspect is inborn. Our species carefully chooses between trust and distrust, as do many other species.

A chimpanzee child, for example, learns to trust its mother. Sitting in a tree, the mother holds it by a hand or foot, dangling it high above the ground. If she'd let go, her child would fall to its death, but no mother would ever do such a thing. During travel, the child clings to her belly, or when it's older, sits on her back. Given a young chimpanzee's dependence of around six years, they in fact *live* on mom's back. I once walked behind a party of chimpanzees in the jungle when a juvenile began standing up and turning somersaults on her mother's back. She got corrected by a brief touch and look-around by her mother, not unlike the way we check rambunctious children in the car's backseat.

Chimpanzee children also learn distrust, for example, when they play with a peer. Both youngsters will make laughing sounds, running around. They seem to enjoy themselves, but there's also a competitive element of who can push whom under and who can hit or be hit without whimpering in pain. Young males, especially, love rough games. But then the big brother of one of them comes around the corner and the whole dynamic changes. The peer whose brother is standing there all of a sudden takes courage, hits harder, or even bites his pal. Both know who will be backed up if it comes to blows. The play has turned into an unfriendly affair. This is nothing unusual: Chimpanzees learn from a very young age that fun lasts only so long as it lasts. No playmate can ever be trusted to the same degree as their mother.

We value trust to an extreme degree. In a group of Bushmen every man may own poison arrows, but they put them point down in a quiver and hang the quiver high up in a tree, out of reach of small children. These arrows are treated the way we would treat live hand grenades. Where would group life be if men were constantly threatening to use them? In our own culture, too, high levels of trust can be found in close-knit communities. Small-town America enjoys so much mutual support and social monitoring that people leave their doors open and cars unlocked. What crimes ever took place in Mayberry?

Large cities are obviously a different story. Think back to 1997, when a Danish mother left her fourteenth-month-old girl in a stroller outside a Manhattan restaurant. Her child was taken into custody and placed in foster care, while the mother ended up in jail. For most Americans, she was either crazy or criminally negligent, but in fact this mother merely did what Danes are used to. Denmark has incredibly low crime rates, and parents feel that what a child needs most is *frisk luft*, or fresh air. The mother counted on safety and good air, whereas New York offered neither. The charges against her were eventually dropped.

When I recently visited Denmark and asked colleagues if they'd leave a child unattended outside, all of them nodded that, yes, this is what everybody does. Not because they don't like to have their babies near them, but because it's so good for them to be out in the open. No one considers the risk of abduction. My hosts were positively puzzled that anyone could consider such a wicked act, and wondered where one would go with an abducted child. If one were to return with it to one's own neighborhood, they surmised, wouldn't everyone ask where that child came from? And wouldn't the connection be quickly made if newspapers were reporting a missing baby?

The faith that Danes unthinkingly place in one another is known as "social capital," which may well be the most precious capital there is. In survey after survey, Danes have the world's highest happiness score.

What Have You Done for Me Lately?

In the same way that my office wouldn't stay empty for long were I to move out, nature's real estate changes hands all the time. Potential homes range from holes drilled by woodpeckers to abandoned burrows. A typical example of a "vacancy chain" is the housing market among hermit crabs. Each crab carries its house around—an empty gastropod shell—in order to cover its behind, literally. The problem is that the crab grows, whereas its house doesn't. Hermit crabs are always on the lookout for new accommodations. The moment they upgrade to a roomier shell, other crabs line up for the vacated one.

It's easy to see supply and demand at work in this hand-me-down economy, but since it plays itself out on a rather impersonal level it remains far removed from human transactions. It would be more interesting if hermit crabs were to strike deals along the lines of "You can have my house, if I can have that dead fish." Hermit crabs are no deal-makers, though, and in fact have no qualms evicting owners by force.

Adam Smith thought that this approach characterized all animals: "Nobody ever saw a dog make a fair and deliberate exchange of one bone for another with another dog..." True, but Smith was wrong to think that animals never trade, or that fairness is alien to them. He underestimated the sociality of animals since he believed that they didn't need one another. My guess, though, is that the great Scottish philosopher would have been delighted to see how far the fledgling field of animal "behavioral economics" has come.

Let me illustrate this with an incident among capuchin monkeys that we taught to pull a tray with food. By making the tray too heavy for a single individual, we gave them a reason to join forces. On this occasion, the pulling was to be done by two females, Bias and Sammy. They successfully brought the food within reach, as they'd done so many times before. Sammy, however, collected her rewards so quickly that she released the tray before Bias had a chance to grab hers. The tray bounced back due to the counterweight and was now out of reach. While Sammy happily munched on her food, Bias threw a tantrum. She screamed her lungs out for half a minute until Sammy approached her pull-bar for the second time, glancing at Bias. She then helped Bias bring in the tray again. This time Sammy didn't pull for her own good, because she only had an empty cup in front of her.

Sammy made this rectification in direct response to Bias's protest against the loss of her reward. Such behavior comes close to human economic transactions: cooperation, communication, and the fulfillment of an expectation, perhaps even an obligation. Sammy seemed sensitive to the quid pro quo: Bias had helped her, so how could she refuse to help Bias? This kind of sensitivity is not surprising given that the group life of these monkeys has the same mixture of cooperation and competition of our own societies.

It is noteworthy that Sammy and Bias are unrelated. The tendency of animals to help kin is obvious in beehives and anthills, but also in mammals and birds. "Blood is thicker than water," we say, and there is plenty of familial support in our own societies as well. Helping kin has genetic advantages, which is why biologists treat it quite separately from helping unrelated individuals. Here the advantages are far less obvious. So, why do animals engage in it? Petr Kropotkin, the Russian prince, offered an explanation at the beginning of the twentieth century in his book *Mutual Aid*. If helping is communal, he argued, all parties stand to gain.

Kropotkin forgot to add that such a system won't work unless everyone contributes more or less equally. Some parties, however, will be tempted to enjoy the fruits of the tree without watering it: In other words, cooperation is vulnerable to freeloaders. A few years after publication of *Mutual Aid*, Kropotkin corrected himself and did mention "loafers," suggesting a solution:

> Let us take a group of volunteers, combining for some particular enterprise. Having its success at heart, they all work with a will, save one of the associates, who is frequently absent from his post. Must they on his account dissolve the group, elect a president to impose fines, or maybe distribute markers for work done? It is evident that neither the one nor the other will be done, but that some day the comrade who imperils their enterprise will be told: "Friend, we should like to work with you; but as you are often absent from your post, and you do your work negligently, we must part. Go and find other comrades who will put up with your indifference!"

Similarly, animals are selective about whom they work with, sometimes developing a "buddy system" of mutually positive partnerships. It's not my intention to be gruesome about it, but a prime example is the way vampire bats share blood night after night depending on each individual's success extracting it from unsuspecting victims. The latter are often large mammals, such as cows or donkeys, but occasionally a sleeping human. Between two buddies, bat A may have been lucky one night and, upon return to the communal roost, regurgitated blood for B. The next night, bat B may have been lucky and did the same for A. Vampire bats can't go one day without blood, so they spread the risk this way.

Chimps go beyond this, however. Males chase after monkeys, which is an extremely challenging mission in three-dimensional space. Since hunters are more successful as a team than alone, they fit the stag-hunt pattern. I once witnessed this spectacle live, which was fascinating except for the "fieldworker's shower" I got due to the fact that chimps defecate profusely when excited. We had learned about the hunt through hooting and screaming mixed with the panicky shrieks of the monkeys. I found myself under a tree in which many adult males had gathered around the carcass of their favorite prey, a red colobus monkey. I am not complaining about the smelly state I found myself in, because it was a real thrill to watch all of this and follow the division of meat. The males shared among themselves as well as with a couple of fertile females.

One hunter usually captures the prey, and not everyone necessarily gets a piece. A male's chance of getting a share appears to depend on his role in the hunt, which hints at reciprocity. Even the most dominant male, if he failed to partake in the hunt, may beg in vain.

Reciprocity can be explored in captivity by handing one chimp a large amount of food, such as a watermelon or leafy branch, and then observing what follows. As if to illustrate Reaganomics, the owner will be center stage, with a cluster of others around him or her, soon to be followed by spin-off clusters around those who obtained a sizable share, until all food has trickled down to everybody. Beggars may whimper and whine, but aggressive confrontations are rare. The few times that they do occur, it's the possessor who tries to make someone leave the circle. She'll whack them over their head with her branch or bark at them in a shrill voice until they leave her alone. Whatever their rank, possessors control the food flow. Once chimps enter reciprocity mode, their social hierarchy takes a backseat.

As in the 1980s hit song "What Have You Done for Me Lately?" the chimps seem to recall previous favors, such as grooming. We analyzed no less than seven thousand approaches to food owners to see which ones met with success. During the mornings before every feeding session, we had recorded spontaneous grooming. We then compared the flow of both "currencies": food and grooming. If the top male, Socko, had groomed May, for example, his chances of getting a few branches from her in the afternoon were greatly improved. We found this effect all over the colony: Apparently, one good turn deserves another. This kind of exchange must rest on memory of previous events combined with a psychological mechanism that we call "gratitude," that is, warm feelings toward someone whose act of kindness we recall. Interestingly, the tendency to return favors was not equal for all relationships. Between good friends, who spend a lot of time together, a single grooming session carried little weight. They both groom and share a lot, probably without keeping careful track. Only in the more distant relationships did small favors stand out and were specifically rewarded. Since Socko and May were not particularly close, Socko's grooming was duly noticed.

The same distinction is found in human society. When, at a conference on reciprocity, a senior scientist revealed that he kept track on a computer spreadsheet of what he had done for his wife and what she had done for him, we were dumbfounded. This couldn't be right. The fact that this was his third wife, and that he's now married to his fifth, suggests that keeping careful score is perhaps not for close relationships. Spouses do obviously maintain a measure of reciprocity, but mutual benefits are supposed to work themselves out over the long haul, not in a tit-for-tat fashion. In close and intimate relationships we prefer to operate on the basis of attachment and trust, reserving our marvelous accounting abilities for relationships with colleagues, neighbors, friends of friends, and so on.

We realize the tit for tat of distant relationships most acutely when it is being violated. Here's a true story of something that happened to my brother-in-law. J. lives in a small seaside town in France, where he is known as the handyman he is. He can build an entire house with his own hands, as he is skilled in carpentry, plumbing, masonry, roof work, and so on. He demonstrates this every day at his own home, and so people naturally ask him for help. Being extremely nice, J. usually dispenses advice or lends a helping hand. One neighbor, whom he barely knew, kept asking about how to put a skylight in his roof. J. lent him his ladder for the job, but since the man kept returning, he promised to come take a look one day.

J. spent from morning until late evening with the neighbor, basically doing the job on his own (as the neighbor could barely hold a hammer, he said), during which time the neighbor's wife came, cooked, and ate lunch (the main meal in France) with her husband without offering J. anything. By the end of the day, he had successfully put the skylight in, having provided expert labor that normally would have cost more than six hundred euros. J. asked for nothing, but when the same neighbor a few days later talked about a scuba diving course, and how it would be fun to do together, he felt this opened a perfect occasion for a return gift, since the course cost about 150 euros. So J. said he'd love to go, but unfortunately didn't have the money in his budget. By now you can guess: The man went alone.

Stories like this make us uncomfortable, perhaps even angry. We closely watch reciprocity, and rightly so, given that it's a core principle of society. Most of it takes place in an unspoken manner, because calling attention to past favors is seen as rude. Although the academic literature on this topic touts the human tendency to punish "cheaters"—who, like this neighbor, take more than they give—in real life punishment is rare. What could J. have done? Thrown a stone through the skylight? Punishment does occur when strangers play games in a psychology lab, but in a small town where everybody knows everybody and stays for years, sometimes generations, one has to choose one's spiteful acts carefully. The only typically human option open to J. is that he could start gossip about what had happened.

But an even simpler solution is to avoid those who are short on gratitude. If one can choose between multiple partners, why not just go with the good ones, who can be trusted to respect past exchanges, rather than those lousy freeloaders whom we can all do without? We are like the clientele of a market where we pick and choose our partnerships, squeezing and smelling them the way the French do with cantaloupes. We want the best ones. People like J.'s neighbor will be out of luck.

This relates to the idea of a "marketplace of services," which I proposed for chimpanzees long ago. The principle of supply and demand reigns everywhere, from bees and flowers (the ratio between flowers and bees determines how attractive the former need to be to get pollinated by the latter) to the curious case of baboons and their babies. Female baboons are irresistibly attracted to infants, and not only their own, but also those of others. They give friendly grunts and like to touch them. Mothers are protective, though, and reluctant to let anyone handle their precious newborns. Interested females groom the mother while peeking over her shoulder or underneath her arm at the baby. After a relaxing session, a mother may permit the other a closer look. The other thus "buys" infant time. Market theory predicts that the value of babies goes up if there are fewer around. In a study of wild chacma baboons in South Africa, Peter Henzi and Louise Barrett found indeed that the length of time mothers were groomed was inversely related to the baby supply in the troop: Mothers of rare infants extracted a considerably higher price than mothers in a troop that had just enjoyed a baby boom.

The way primates trade commodities mimics our economies in surprising ways. We even managed to set up a miniature "labor market" among capuchin monkeys, which was inspired by their natural hunting behavior. These monkeys can be true cooperators, such as when they encircle giant squirrels, which are as hard to catch as monkeys are for chimpanzees. Afterward, the capuchins share the meat.

This situation can be modeled by requiring parties to collaborate but rewarding only one of them. In a variation on the cooperative pulling described for Sammy and Bias, we set up tests in which only one puller received a cup filled with apple slices. We called him the "CEO." His partner had an empty cup in front of him, so pulled for the CEO's benefit. We called this individual the "laborer." Both monkeys sat side by side, separated by mesh, and could see both cups. From previous tests we knew that food possessors often bring food to the mesh, where they permit their neighbor to reach for it. On rare occasions they'd push pieces to the other.

As it turned out, if both monkeys brought in the food, the CEO shared more through the mesh with the other than if he had secured the food by himself. In other words, he shared more after having been helped. We also found that sharing fosters cooperation, because if the CEO was stingy with his food, the success rate of the pair dropped dramatically. Without sufficient rewards, laborers simply went on strike.

In short, monkeys seem to connect effort with compensation. Perhaps owing to their collective pursuits in the wild, they grasp the first rule of the stag hunt, which is that joint effort requires joint rewards.

Evolution Sans Animals

It's unknown how long monkeys keep favors in mind. Their reciprocity may be merely "attitudinal" in that they mirror immediate attitudes. If others are hostile, they'll be hostile back. If others are nice, they'll be nice back. Consequently, if another monkey helps them pull a heavy tray, they'll share in return.

Hare Krishna followers bet on this principle when they hand out flowers to pedestrians. As soon as a flower is accepted, they ask for money. Instead of simply begging for a handout, they count on the fact that we mirror their behavior. We apply it every day in fleeting contacts, such as with people whom we meet on the train, at parties, at sports games, and so on. Since attitudinal reciprocity doesn't require record keeping, it isn't mentally taxing.

But, like us, some animals follow a more complex scheme, storing favors in long-term memory. In our food-for-grooming experiment, chimps did so for at least a couple of hours, but I have known apes who've been grateful for years. One was a female whom I had patiently taught to bottle-feed an adopted infant. Previously, she had lost several offspring due to insufficient lactation. Chimps being tool users, she had no trouble handling a nursing bottle. In the ensuing years, this female raised her own infants this way as well. Decades later, she was still thrilled if I stopped by the zoo where she lived. She'd groom me with enthusiastic tooth-clacking, showing that I was a hero to her. Animal keepers, most of whom were unaware of our history, couldn't believe the fuss she was making over me. I'm convinced it had to do with me having helped her overcome a problem that had given her unimaginable grief.

If chimps look back further than monkeys, remembering previous events more clearly, this makes their reciprocity more deliberate and calculated. If a wild chimp, for example, removes a poacher's snare that has tightened around the wrist of another—having caused the other to scream in excruciating pain—it's safe to assume that his assistance will be remembered. It's even possible that chimps not only look back, but also forward, treating others nicely so as to curry favors. I can't say that this has been proven, but evidence is mounting. For example, when male chimps vie for high status, they try to make friends with as many potential backers as possible. They do the rounds with females, grooming them and tickling their offspring. Normally, male chimps are not particularly interested in the young, but when they need group support they can't stay away from them. Do they know that all female eyes rest on them to see how they treat the most vulnerable?

The tactic is eerily humanlike. I regularly download pictures of American politicians holding up babies under the eyes of their parents, who look on with a mixture of delight and apprehension. Have you ever noticed how often politicians lift infants above the crowd? It's an odd way of handling them, not always enjoyed by the object of attention itself. But what good is a display that stays unnoticed?

And then there are the occasions when a political contender, such as a rising young chimpanzee male, becomes extraordinarily generous with females around the time that he begins to challenge the leader. In a process that may take months, the contender irritates the established alpha multiple

Political candidates love to hold babies up in the air.

times a day to see what kind of reaction he gets. At the same time, he shares food specifically with those who might assist him in his quest. At the Arnhem Zoo, I saw rising males go out of their way to secure goodies: braving electric wire around live trees, jumping over it, to climb up to the foliage and break off branches for the mass gathered below. Such behavior seemed to boost their popularity.

In the wild, high-ranking male chimps are said to bribe others, sharing meat selectively with potential allies and withholding it from their rivals. And at Bossou, in Guinea, male chimps customarily raid surrounding papaya plantations—a perilous undertaking—and bring the delicious fruits back to buy sex with: They specifically share with fertile females. According to British scientist Kimberley Hockings, "Such daring behavior certainly seems to be an attractive trait and possessing a sought-after food item, such as papaya, appears to draw positive attention from the females."

It isn't exactly the bone-trading among dogs envisioned by Smith, but we're getting close. Chimps may have foresight along the lines of "If I do this for him or her, I may get that in return." Such calculations would explain observations at Chester Zoo in the United Kingdom. During fights in its large chimpanzee colony, individuals enjoyed support from parties whom they had groomed the day before. Not only this, but they seemed to plan whom they'd pick a fight with, grooming potential supporters days in advance so that the outcome might turn in their favor.

Given the elaborate exchanges among our close relatives, perhaps even including planning and foresight, one wonders why some students of human reciprocity define their field in opposition to animal behavior. They call human cooperation a "huge anomaly" in the natural world. It's not that the followers of this school are anti-evolutionary—on the contrary, they are self-proclaimed Darwinists—but they are eager to keep hairy creatures on the sidelines. I have only half-jokingly called their approach "evolution sans animals." They have been quick to write off chimpanzee cooperation as a product of genetic kinship, thus putting it in the same category as the communal life of ants and bees. Only humans, they say, engage in large-scale co-operation with nonrelatives.

When zoo studies made clear that kinship is not required for chimps to work closely together, this was dismissed as not represent-ing the natural condition. And when wild apes, too, were shown to regularly cooperate with unrelated individuals, this was questioned, because isn't it hard to know exactly who's related to whom? Can we really exclude the possibility that males who have formed a coalition are brothers or cousins? Nothing could convince the skeptics. Even-tually, this fruitless debate was settled by new technology. It's not un-usual nowadays for primatologists to return from the field with a load of carefully labeled fecal samples. DNA extraction from these samples offers a more accurate picture of genetic relations than ever before, telling us which individuals are related to the nth degree, which male fathered which offspring, who immigrated into the community from the outside, and so on.

One of the most complex field projects was set up in Kibale Na-tional Park in Uganda, which combined years of data on chimpanzee social behavior with excrement picked up from the forest floor. It's hard to imagine how much sweaty, smelly work goes into a genotyp-ing project like this, but the results were more than worth it. First of all, the German-American team demonstrated that kinship matters: Broth-ers spend more time together, support one another more, and share more food than unrelated males. This is of course exactly what one would expect, not only in chimpanzees but also in any small-scale human society. But the study also demonstrated widespread coopera-tion among nonrelatives. In fact, the majority of close partnerships in the Kibale community were between males lacking family ties.

This suggests mutualism and reciprocity as the basis of coopera-tion, thus placing chimps much closer to humans than to the social insects. Nothing surprising there, but it also means that to understand the psychology of human reciprocity, apes offer a perfect comparison. That is not to deny a few differences with human cooperation, one of which may be a more developed tendency in our species to penalize those who fall short. But even this difference may be less absolute than it sounds. We know that chimps get even with those who have turned against them. Hours after an incident in which others banded together against him, a high-ranking male may seek out his tormentors individually, while they are sitting somewhere alone, and teach them a lesson they won't forget. Chimps settle scores just as easily as they return favors, so I wouldn't put it past them to impose sanctions on others.

My guess is that humans show all of these tendencies to a greater extent, and thus are capable of more complex, larger-scale cooperation. If hundreds of workers build a jet airliner, all relying on one another, or if many different levels of employees make up a company, this is possible only because of our advanced abilities of organization, task division, storing of past interactions, connecting effort with reward, building trust, and discouraging freeloading. Human psychology evolved to permit ever larger and more complex stag hunts, going well beyond anything in the animal kingdom. While the actual hunting of large prey may have driven this evolution, our ancestors engaged in other cooperative ventures, such as communal care for the young, warfare, the building of bridges, and protection against predators. They benefited from cooperation in myriad ways.

One school of thought proposes that our ancestors became such great team players because of their dealings with strangers. This forced them to develop reward and punishment schemes that worked even with outsiders whom they had never met and would never see again. It is well known that human strangers brought together in the laboratory adopt strict rules of cooperation and turn against anyone who fails to comply. This is known as *strong reciprocity*. We just get very upset if we put in a lot of effort and then get shortchanged by someone who acts as if he's playing along but in fact takes advantage of us. We have all kinds of ways to exclude or punish such people. But while no one doubts that we disapprove of cheaters, the evolutionary origin of these feelings is a point of debate. The fact that we apply norms to strangers doesn't necessarily mean that these norms evolved specifically for this purpose. Were strangers really that important in human evolution? Robert Trivers, the originator of the theory of reciprocal altruism, doubts it:

> If humans show strong dispositions towards fairness in one-shot, anonymous encounters, this hardly means that these dispositions evolved to function in one-shot, anonymous encounters any more than we would argue that children's strong emotional reactions to cartoons show that such reactions evolved in the context of cartoons.

Remember our discussion of "motivational autonomy," of how a behavior may have evolved for reason X, yet in reality be used for reasons X, Y, and Z? The example I gave was parental care, which evolved for the benefit of offspring yet is often applied to adopted children, even household pets. In the same way, Trivers believes that norms of exchange began between individuals who knew one another and lived together, after which they were extended to strangers. We shouldn't focus too much on anonymous encounters, therefore, because the true cradle of cooperation is the community. This is of course also the context in which apes engage in social exchange, so that the difference with humans is probably less dramatic than originally thought.

In fact, evolution never produces "huge anomalies." Even the neck of the giraffe is still a neck. Nature knows only variations on themes. The same applies to cooperation. Trying to set human cooperation apart from the larger natural scheme including apes, monkeys, vampire bats, and cleaner fish hardly qualifies as an evolutionary approach.

The Last Shall Be First

How often have you seen rich people march in the street shouting that they're earning too much? Or stockbrokers complaining about the "Onus of the Bonus!"? The well-to-do rather follow Bob Dylan's observation that "man is opposed to fair play, he wants it all and he wants it his way." Instead, protesters typically are blue-collar workers yelling that the minimum wage has to go up, or that their jobs shouldn't go overseas. A more exotic example was the 2008 march by hundreds of women through the capital of Swaziland. Given their destitute economy, they felt that the king's wives had overstepped their privileges by chartering an airplane for a shopping spree in Europe.

Fairness is viewed differently by the haves and have-nots. The reason for stating the obvious is the common claim that our sense of fairness transcends self-interest, that it's concerned with something larger than ourselves. True, most of us subscribe to this ideal, as do many of our institutions. Yet it's also clear that this is not how fairness started. The underlying emotions and desires aren't half as lofty as the ideal itself. The most recognizable emotion is resentment. Look at how children react to the slightest discrepancy in the size of their pizza slice compared with their sibling's. They shout "That's not fair!" but never in a way transcending their own desires. As a matter of fact, in my younger years I've had fights like this with my wife until we hit on the brilliant solution that one of us would do the dividing and the other the choosing. It's amazing how quickly one develops perfect cutting skills.

We're all for fair play so long as it helps us. There's even a biblical parable about this, in which the owner of a vineyard rounds up laborers at different times of the day. Early in the morning, he goes out to find men, offering each one a denarius for their labor. He goes again in the middle of the day, offering the same. At the "eleventh hour" he hires a few more with the same deal. By the end of the day, he pays all of them, starting with the last ones hired. Each one gets a denarius. Watching this, the other workers expect to get more since they had worked through the heat of the day. Yet they get paid one denarius as well. The owner doesn't feel he owes them any more than what he had promised. The passage famously concludes with "So the last will be first, and the first will be last."

Again, the grumbling was one-sided: It came from the early hires.

They gave the master the "evil eye," whereas those who had worked the least didn't give a peep. The only reason the latter might have been unhappy is that the situation obviously didn't endear them to the others. They'd be wise not to gloat and celebrate. The potential of green-eyed reactions is the chief reason why we strive for fairness even when we have the advantage. To my own amazement, I find myself siding here with the "Monster of Malmesbury," as Thomas Hobbes was known, when he stated that we're interested in justice only for peace's sake.

Am I uncharacteristically cynical? You've heard me explain at length how incredibly empathic, altruistic, and cooperative we are, so why when we get to fairness is self-interest all that I see? The inconsistency isn't as great as it may seem, because I do believe that all human (or animal) behavior must in the end serve the actors. In the domain of empathy and sympathy, evolution has created a stand-alone mechanism that works whether or not our direct interests are at stake. We are driven to empathize with others in an automated, often unconditional fashion. We genuinely care about others, wanting to see them happy and healthy regardless of what immediate good this may do for us. We evolved to be this way because, on average and in the long run, it served our ancestors. But I fail to see how the same applies to our sense of fairness. Other-orientation seems such a small part of it. The chief emotions are egocentric, preoccupied with what we get compared to others, and how we may come across to others (we do like to be seen as fair-minded). Only secondarily is there an actual concern for others, mostly because we long for a livable, harmonious society. The latter desire is also visible in other primates when they break up disputes in their midst or bring conflicted parties together. We have gone one step further, though, by being sensitive to how resource distribution impacts everybody around us.

The reaction of children to perceived unfairness shows how deeply seated these sentiments are, and the egalitarianism of hunter-gatherers suggests its long history. In some cultures, hunters aren't even allowed to carve up their own kill, so as to prevent them from favoring their family. The antiquity of fairness is underappreciated by those who regard it as a noble principle of recent origin, formulated by wise men during the French Enlightenment. I seriously doubt that we will ever appreciate the human condition by looking back a couple of centuries rather than millions of years. Do wise men ever formulate anything new, or are they just good at reformulating what everybody knows? They often do so in an admirable fashion, but to credit them with the concepts themselves is a bit like saying that the Greeks invented democracy. The elders of many preliterate tribes listen for hours, sometimes days, to the opinions of all members before making an important decision. Aren't they democratic? Similarly, the fairness principle has been around since our ancestors first had to divide the spoils of joint action.

Researchers have tested this principle by offering players an opportunity to share money. The players get to do this only once. One player is given the task to split the money into two—one part for himself, the remainder for his partner—and then propose this split to the other. It is known as the "ultimatum game," because as soon as the offer has been made, the power shifts to the partner. If he turns down the split, the money will be gone and both players will end up empty-handed.

If humans are profit maximizers, they should of course accept *any* offer, even the smallest one. If the first player were to give away, say, $1 while keeping $9 for himself, the second player should simply go along. After all, one dollar is better than nothing. Refusal of the split would be irrational, yet this is the typical reaction to a 9:1 split. A comparison of fifteen small-scale societies by American anthropologist Joseph Henrich and his team found some cultures to be fairer than others. Offers in these faraway places (in local currency, and sometimes using tobacco instead of money) ranged from an average of $8 for the first player and $2 for the second all the way to $4 for the first and $6 for the second. Even the latter, hypergenerous offers were rejected in cultures in which making a large gift is a way of making others feel inferior. For most cultures, however, offers were close to equal, often with a slight advantage for the first player, such as a $6 versus $4 division. This is also the typical offer in modern societies, such as when university students play the ultimatum game.

Fairness is understood all over the world, including places untouched by the French Enlightenment. Players avoid highly skewed proposals. That they don't want to appear greedy is understandable: Brain scans of players facing unfair proposals reveal negative emotions, such as scorn and anger. The beauty of the ultimatum game is, of course, that it offers an outlet for these feelings. Those who feel slighted can punish the proposer even though in doing so they also punish themselves.

That we're willing to do so shows that certain goals take priority over income. We know an unfair distribution when we see one, and try to counteract it. That this is mainly done for the sake of good relations explains why the above multicultural study measured the fairest offers in societies with the highest levels of cooperation. A good example is the Lamalera whale hunters, in Indonesia, who capture

Lamalera whale hunting is a dangerous enterprise with huge payoffs—the sort of collective activity that puts a premium on fair distribution.

whales almost bare-handed. Entire families are tied together around an extremely dangerous activity on the open ocean by a dozen men in a large canoe. Since these men are literally in the same boat, a fair distribution of the food bonanza is very much on their mind. In contrast, societies with greater self-sufficiency, in which every family takes care of itself, are marked by unfair offers in the ultimatum game. It's easy to recognize the stag-versus-hare-hunt scenarios here: Human fairness goes hand in hand with communal survival.

That this connection may be quite ancient became clear when a student, Sarah Brosnan, and I discovered it in monkeys. While testing capuchin monkeys in pairs, Sarah had noticed how much they disliked seeing their partner get a better reward. At first, this was just an impression based on their refusal to participate in our tests. We weren't too surprised. But then we realized that economists had given these reactions the fancy label of "inequity aversion," which they had turned into a topic of serious academic debate. This debate obviously revolved around human behavior, but what if monkeys showed the same aversion?

Testing two monkeys at a time, Sarah would offer a pebble to one and then hold out her hand so that the monkey could give it back in exchange for a cucumber slice. Alternating between them, both monkeys would happily barter twenty-five times in a row. The atmosphere turned sour, however, as soon as we introduced inequity. One monkey would still get cucumber, while its partner now enjoyed grapes, a favorite food. The advantaged monkey obviously had no problem, but the one still working for cucumber would lose interest. Worse, seeing its partner with juicy grapes, this monkey would get agitated, hurl the pebbles out of the test chamber, sometimes even throwing those paltry cucumber slices. A food normally devoured with gusto had become distasteful.

Discarding perfectly fine food simply because someone else is getting something better resembles the way we reject an unfair share of money or grumble about an agreed-upon denarius. Where do these reactions come from? They probably evolved in the service of cooperation. Caring about what others get may seem petty and irrational, but in the long run it keeps one from being taken advantage of. It's in everyone's interest to discourage exploitation and free riding, and make sure that one's interests are taken seriously. Our study was the first to show that these reactions may have been around for as long as animals have engaged in tit for tat.

Had Sarah and I merely spoken of "resentment" or "envy," our findings might have gone unnoticed. But because we saw no reason why our monkeys weren't showing inequity aversion, we drew the keen, somewhat baffled interest of philosophers, anthropologists, and economists, who almost choked on the comparison with monkeys. Indignant commentaries in prestigious journals followed, as well as a sharp jump in speaking invitations. As it happened, our study came out the day Richard Grasso, head of the New York Stock Exchange, was forced to resign because of a public outcry over a pay package of close to $200 million. Commentators couldn't resist contrasting the unbridled greed in human society with our monkeys, suggesting that we could learn a thing or two from them.

I had to think back to this comparison when, in 2008, the U.S. government proposed a huge bailout of the financial industry. The unimaginable number of tax dollars combined with deep resentment of "fat cat" CEOs who had gambled so much money away led to public fury in the media. As one business magazine put it, "Underlying public distrust of the wealthy and the perception the $700 billion mortgage bailout will help big banks and rich CEOs continues to be the main stumbling block and minefield for passage of a rescue package." Some saw this bailout as the end of laissez-faire economics, comparing the blow it dealt to capitalism with the way communism had been impacted by the fall of the Berlin Wall. But for me the more interesting part was how people reacted, such as the obvious Schadenfreude about CEOs demoted from their extravagant corner office to a less ostentatious location. When this happened to Richard Fuld, the chief of Lehman Brothers, artist Geoffrey Raymond created *The Annotated Fuld*, a painted portrait on which fired employees could leave farewell scribbles. Needless to say, they didn't show much love for their multimillion-dollar boss, with comments like "Bloodsucker!," "Greed!," and "I hope his villa is safe!"

If possible, the reactions were even worse when it emerged that some companies had enjoyed luxurious retreats at the same time that they were negotiating for government help. One company had sent its executives to a fancy spa, with massages and all. Another had organized a partridge hunt in England where its employees walked around in tweed knickers, and sipped fine wine at lavish feasts. One executive told an undercover reporter, "The recession will go on until about 2011, but·the shooting was great today and we are relaxing fine." A month later, Detroit's Big Three automakers arrived in Washington to plead for financial support of their ill-managed industries and faced an outcry when the public learned that each CEO had flown in on his own private jet. Hadn't they noticed how much the country was fed up with excess at the top? The always subtle columnist Maureen Dowd exclaimed: "Heads must roll."

Despite the obvious parallels between this outrage and primate behavior, it's nevertheless useful to point out what our monkeys' reaction was *not*. We can exclude both the simplest and most complex explanations. At the simple end, one could argue that the sight of grapes lessens the appeal of cucumber the same way that most men won't touch a glass of water if you put a beer next to it. In other words, it was not so much what their partner was getting that ticked off our monkeys, but that they were simply holding out for something better. To test this, we added a twist to our study. Before each equity test, in which both monkeys ate cucumber, we'd wave grapes around, just to show that we had them. This may seem cruel, but it hardly bothered the monkeys: They still contentedly traded for cucumber. Only if the grapes were actually given to their partner did the one who missed out go into protest mode. It really was the inequity that bothered them.

At the more complex end, our monkeys did not seem to follow a fairness *norm*. A norm applies equally to everyone, which would mean in this case that the monkeys were not only concerned about getting less but also about getting more than others. There was no evidence for the latter, however. The monkey with the advantage, for example, never gave away any of her grapes so as to equalize the distribution. If we speak of "fairness," therefore, it should be understood as the most egocentric kind, similar to the treatment that young children spill tears over.

This applies to monkeys. For apes, on the other hand, we cannot rule out a fairness norm. They seem to monitor their interactions more closely and keep better track of each individual's contributions to common goals. Chimpanzees, for example, regularly break up fights over food without taking any of it. I once saw an adolescent female interrupt a quarrel between two youngsters over a leafy branch. She took the branch away from them, broke it in two, then handed each one a part. Did she just want to stop the fight, or did she understand something about distribution? There's even one observation of a bonobo worried about getting too much. While being tested in a cognitive laboratory, a female received plenty of milk and raisins but felt the eyes of her friends on her, who were watching from a distance. After a while, she refused all rewards. Looking at the experimenter, she kept gesturing to the others until they too got some of the goodies. Only then did she finish hers.

This bonobo was doing the smart thing. Apes think ahead, and had she eaten her fill right in front of the rest, there might have been repercussions when she rejoined them later in the day. Privileges are always enjoyed under a cloud. Human history is filled with "let them eat cake" moments that create resentment, sometimes boiling over into bloody revolt. I cannot help but look through the same lens at a gruesome chimpanzee attack on a human, which curiously revolved around cake as well. The central figure was Moe, well-known to the media for a string of incidents in his long career as pet chimpanzee. The last time he was in the news concerned his escape, in 2008, from a California sanctuary surrounded by mountains covered with thick brush. Except for one unconfirmed "monkey" sighting at a nearby nudist camp, and massive efforts with helicopters, bloodhounds, and surveillance cameras, Moe was never seen again.

Moe had been brought from Africa as a baby and lovingly reared by an American couple who treated him as their child for as long as they could. But apes are too strong and wily to make good pets. The couple was forced to move Moe to a sanctuary after he attacked a woman and a police officer. They regularly visited "their boy" at the sanctuary. A few years before his final escape, on Moe's thirty-ninth birthday, they brought him a load of sweets to celebrate. Moe received a magnificent raspberry cake, drinks, and new toys, which would all have been fine if there had been no other chimps around. But this was not the case: The sanctuary had taken in other chimps from abusive homes and Hollywood trainers. While Moe was feasting on his cake under the eyes of his foster parents, two male chimps in another enclosure managed to break out. They went straight for the husband. My guess is that they would have attacked Moe if he hadn't been behind bars. Even though this incident has gone down as one of the most horrific animal assaults ever on a human, it is in line with how male chimps attack members of their own species. The two chimps chewed off most of the man's nose, face, and buttocks, tore off his foot, and bit off both testicles. He was lucky to survive, which only happened because his attackers were shot.

It is unclear if the motive for the assault was territorial (chimps don't take kindly to strangers) or rather had to do with all of the attention and goodies lavished upon Moe. The inequity of this unintended experiment exceeded anything ever introduced in our studies. If monkeys get upset by having to make do with cucumbers while others eat grapes, you can imagine how chimps react to seeing one of them own the candy store. Moe's owners probably hadn't realized how sensitive chimps are to unequal treatment, especially if the advantaged one isn't even a friend.

The main reason humans seek fairness, I believe, is to prevent such negative reactions. Even the Monster of Malmesbury thought so, as did the "Sage of Baltimore," H. L. Mencken, who said, "If you want peace, work for justice." This is not to deny a role for other-regarding feelings. The golden rule is universally appreciated, and most of us reach a point at which we genuinely feel that others deserve the same treatment that we like for ourselves. We easily produce this rationalization for fairness, which definitely adds power to it, but deep down we also realize what's at stake. Whatever noble reasons we give for fairness and justice, they have the firm backing of our vested interest in a harmonious and productive social environment.

Other primates seem to adopt a narrower view, focused on immediate benefits, yet it's too early to conclude that they don't have a fairness norm. Studies on inequity aversion in animals have only just begun. When Sarah and I tested chimps on grape-versus-cucumber deals, we found reactions similar to those of the capuchin monkeys. But we also explored another well-known human tendency, which is that we relax the rules in close relations. Between family, friends, and spouses we don't keep as careful track of favors and inequities as we do with acquaintances, neighbors, and colleagues. The chimp data confirmed this difference. Individuals who had spent little time together (similar to Moe and the other sanctuary chimps) showed by far the strongest reaction to getting the short end of the stick, whereas the members of a colony established thirty years ago hardly blinked. Having played together while young and having grown up together, these chimps were virtually immune to inequity. Social closeness apparently makes apes, like humans, less touchy about this issue.

Inequity aversion will no doubt prove a rich area of research, all the more so since there is no reason to think it's limited to primates. I expect it in all social animals. A most entertaining account concerns Irene Pepperberg's typical dinner conversation with two squabbling African gray parrots, the late Alex and his junior colleague, Griffin:

> I then had dinner, with Alex and Griffin as company. Dining company, really, because they insisted on sharing my food. They loved green beans and broccoli. My job was to make sure it was equal shares, otherwise there would be loud complaints. "Green bean," Alex would yell if he thought Griffin had had one too many. Same with Griffin.

Another species in which to expect such reactions is the domestic dog, which descends from cooperative hunters used to dividing prey. At the Clever Dog Lab at the University of Vienna, Friederike Range found that dogs refuse to lift their paw for a "shake" with a human if they get nothing for it while their companion is rewarded. Disobedient dogs show signs of tension, such as scratching and looking away. The reward itself isn't the issue, because the same dogs are perfectly willing to obey if *neither* one receives food. So dogs too may be sensitive to injustice.

Monkey Money

In the 1930s, when the Yerkes National Primate Research Center was still located in Orange Park, Florida, scientists decided to introduce apes to the wonders of money. They rewarded them with poker chips to be used in a "chimpomat": a vending machine that delivered food upon insertion of a token. The chimps first needed to understand that the chips were promissory notes to be accumulated and converted. After they had learned this, the scientists introduced chips of different value, such as a white chip worth one grape and a blue one worth two grapes. The chimps quickly learned to prefer the highest-value chips.

Our capuchin monkeys, too, have learned to use tokens in exchange for goodies. In one study, Sarah even got them to learn from one another. One monkey bartered with two kinds of tokens, getting bell pepper pieces for one kind and Froot Loops sweet cereal for the other. Bell peppers rank near the bottom of the preference scale, whereas Froot Loops rank near the top. Just from watching the proceedings, a monkey sitting next to the exchanger would develop a preference for tokens that offered the best deal.

We exploited these monetary skills in the experiment described a few chapters back, in which one capuchin chose between a "selfish" token that rewarded only itself and a "prosocial" token that rewarded both itself and a partner. Our monkeys overwhelmingly preferred the prosocial option, thus demonstrating that they care for one another.

This is also well-known for chimps, both in the way they help one another defeat rivals, console distressed parties, defend one another against leopards, and show targeted helping in experiments. Prosociality has a long evolutionary history.

Nevertheless, egoism always lurks around the corner. While testing capuchins with selfish versus prosocial options, we found three ways in which we could kill their tendency to be nice. The first is to pair them with a stranger: They are in a much more selfish mood with partners that they've never met before. This fits the idea of the in-group as the cradle of cooperation.

One capuchin monkey reaches through an arm hole to choose between differently marked pieces of pipe while another looks on. The pipe pieces can be exchanged for food. One token feeds both monkeys; the other feeds only the chooser. Capuchins typically prefer the "prosocial" token.

The second, even more effective way to reduce prosociality is to put the other out of sight by sliding a solid panel between both monkeys. Even if the monkey making the choice knows the one on the other side well, and has seen the other through a small peephole, it still refuses to be prosocial. It acts as if the other isn't there, and turns completely selfish. Apparently, in order to share they need to see their partner. Humans report feeling good while doing good, and brain scans show that our reward centers light up when we give to others. Monkeys may get the same satisfaction from generosity, but only if they can see the outcome, which recalls one of the oldest definitions of human sympathy, according to which we derive pleasure from seeing another's fortune.

Humans have great imagination. We can visualize a poor family wearing the clothes we sent them or children sitting in the school that we helped build at the other end of the globe. Just thinking of these situations makes us feel good. Monkeys probably can't project the effects of their actions across time and space, and so the "warm glow" of giving reaches them only if the beneficiary is in plain view. The emotions involved may not be that different between humans and monkeys, but monkeys express them only under a narrow set of circumstances.

The third way to eliminate acts of kindness is perhaps the most intriguing, since it relates to inequity. If their partner gets a superior reward, our monkeys become reluctant to pick the prosocial option. They are perfectly willing to share, but only if their partner is visible and gets what they get themselves. As soon as their partner is better off, competition kicks in and interferes with generosity.

The same competition can be recruited to squeeze more out of primates in the same way that our economies use it to squeeze more out of people. If you want to keep up with the Joneses, you'll just need to work a bit harder. Pepperberg exploited the competitiveness of her parrots, and we have noticed that our chimps perform better if we give their rewards to others. When a chimp selects images on a touchscreen, for example, he may do so a hundred times in a row. But inevitably his attention wanders, and errors result in less fruit. If, instead of just withholding these rewards, we actually give them to a nearby companion, our subjects suddenly become very keen on the task. They stay with their eyes glued to the screen and apply themselves so as to prevent their goodies from going to the other. We call this the "competitive reward" paradigm.

Given our interest in such competition, you can imagine how puzzled we were by an out-of-the-blue e-mail telling us that we must be "communists," because who else would see fairness as part of human nature? Mind you, we get the strangest e-mails (a recent example: a picture of an abundantly hairy chest sent by a man who felt he had ape ancestry, which we of course couldn't deny), but this particular message sounded rather angry, accusing us of legitimizing social tendencies that our correspondent clearly didn't approve of even in humans. Fairness and justice, what romantic drivel! The funny thing is that the impression we have of our monkeys is the exact opposite. We look at them as little capitalists with prehensile tails, who pay for one another's labor, engage in tit for tat, understand the value of money, and feel offended by unequal treatment. They seem to know the price of everything.

What confuses some is that fairness has two faces. Income equality is one, but the connection between effort and reward is another. Our monkeys are sensitive to both, as are we. Let me explain the difference by contrasting Europe and the United States, which traditionally emphasize different sides of the same fairness coin.

When I first arrived in the United States, I had a mixed impression: On the one hand I felt that the United States was less fair than what I was used to, but on the other hand it was more fair. I saw people living in the kind of poverty that I knew only from the third world. How could the richest nation in the world permit this? It became worse for me when I discovered that poor kids go to poor schools and rich kids to rich schools. Since public schools are financed primarily through state and local taxes, there are huge differences from state to state, city to city, and neighborhood to neighborhood. This contrasts with my own experience, in which all children shared the same school regardless of their background. How can a society claim equal opportunity if the location of one's birth determines the quality of one's education?

But I also noticed that someone who applies him- or herself, as I surely intended to do, can go very far. Nothing stands in their way. Envy is far from absent, and is in fact somewhat of a joke in academia ("Why do academics fight so much? Because there's so little at stake!"), but generally speaking, people are happy for you if you succeed, congratulate you, give you awards, and raise your salary. Success is something to be proud of. What a relief compared to cultures in which the nail that sticks out gets hammered down, or my own country, with its fine Dutch expression, "Act normally, which is crazy enough!"

Holding people back from achievement by hanging the weight of conformity around their necks disrupts the connection between effort and reward. Is it fair for two people to earn the same if their efforts, initiatives, creativity, and talents differ? Doesn't a harder worker deserve to make more? This libertarian fairness ideal is quintessentially American, and feeds the hopes and dreams of every immigrant.

For most Europeans, this ideal takes a backseat to the advice from Dolly Levi, played by Barbra Streisand in the 1969 movie *Hello, Dolly!*, who exclaimed: "Money, pardon the expression, is like manure. It's not worth a thing unless it's spread around." I have seen European newspaper editorials argue that television personalities should never earn more than the head of state, or that CEO salaries should never rise by a greater percentage than worker payment. As a result, Europe is a more livable place. It lacks the giant, nearly illiterate underclass of the United States, which lives on food stamps and relies on hospital emergency rooms for its health care. But Europe also has less of an incentive structure, resulting in a lower motivation for the unemployed to get jobs or for people to start a business. Hence the exodus of young entrepreneurs from France to London and other places.

U.S. CEOs easily earn several hundred times as much as the average worker, and the Gini index (a measure of national income inequality) of the United States has risen to unprecedented heights. The proportion of income owned by the wealthiest 1 percent of Americans recently returned to the level of the Great Depression. The United States has become a winner-take-all society, as Robert Frank called it, with an income gap that seriously threatens its social fabric. The more the poor resent the rich, the more the rich fear the poor and retreat into gated communities. But an even greater burden is health: U.S. life expectancy now ranks below that of at least forty other nations. In principle, this could be due to recent immigration, lack of health insurance, or poor eating habits, but the relation between health and income distribution is in fact not explained by any of these factors. The same relation has also been demonstrated *within* the United States: Less egalitarian states suffer higher mortality.

Richard Wilkinson, the British epidemiologist and health expert who first gathered these statistics, has summarized them in two words: "inequality kills." He believes that income gaps produce social gaps. They tear societies apart by reducing mutual trust, increasing violence, and inducing anxieties that compromise the immune system of both the rich and the poor. Negative effects permeate the entire society:

> It seems that the most likely reason income inequality is related to health is because it serves as a proxy for the scale of social class differentiation in a society. It probably reflects the scale of social distances and the accompanying feelings of superiority and inferiority or disrespect.

Now, don't get me wrong: No one in his right mind would argue that incomes should be leveled across the board, and only the most die-hard conservatives believe that we lack any obligation to the poor. Both kinds of fairness—the one that seeks a level playing field and the one that links rewards to effort—are essential. Both Europe and the United States pay a steep price, albeit different ones, for stressing one fairness ideal at the expense of the other. After having lived for so long in the United States, I find it hard to say which system I prefer. I see the pros and cons of both. But I also see it as a false choice: It's not as if both fairness ideals couldn't be combined. Individual politicians and their parties may be committed to either the left or right side of this equation, but every society zigzags between these poles in search of an equilibrium that offers the best economic prospects while still fitting the national character. Of the three ideals of the French Revolution—liberty, equality, and fraternity—Americans will keep emphasizing the first and Europeans the second, but only the third speaks of inclusion, trust, and community. Morally speaking, fraternity is probably the noblest of the three and impossible to achieve without attention to both others.

Fraternity is also easiest to understand from a primate perspective, with survival relying so heavily on attachment, bonding, and group cohesion. Primates evolved to be community builders. Nevertheless, they are no stranger to equalizing tendencies and the link between effort and reward. When Bias screamed at Sammy for letting her food get out of reach, she was protesting the loss of rewards that she had *worked* for. This was not just about equality. Like the vineyard workers, Bias seemed to take effort into account. In fact, in one of our studies we found that the more effort it takes to earn rewards, the more sensitive a primate becomes to seeing another get something better. It's as if they're saying, "After all this work, I still don't get what he gets?!"

Such reactions typify primates with egalitarian tendencies but don't necessarily apply to those that are strictly hierarchical, such as baboons. Baboons are marked by low social tolerance and empathy. When American primatologist Benjamin Beck watched a female baboon assist a male at the Brookfield Zoo near Chicago, his account offered an interesting reflection on dominance. Baboon males are twice the size of females and possess daggerlike canine teeth, and thus there's never any doubt about who outranks whom. A female, named Pat, had learned to pick up a long rod in another part of the cage that was inaccessible to the male, Peewee. Peewee, in turn, knew how to use the rod to pull in food. Previously he had used the tool on his own, sharing only scattered bits of food with Pat. But the first time Pat spontaneously fetched his tool, which she did after a long grooming bout between the two, Peewee became like a new baboon. Having collected the bounty, he shared fifty/fifty with Pat. It was as if he recognized her contribution. But the more their cooperation grew, the more Pat's share dwindled. In the end, she had to content herself with only about 15 percent. This was still better than nothing, which may explain why she kept bringing the rod, but it's the sort of share that humans soundly reject in the ultimatum game. And not only humans: Had Pat been a capuchin monkey, or a chimpanzee, she would have thrown screaming tantrums at her compensation package.

I cannot help but ponder all of these fine distinctions between rank, equality, inequality, and deserved versus undeserved payoffs while reading passages, as in *Suite Française*, about aristocrats mingling with the common people. The context of an industrialized multilayered society is new but the emotional undercurrent of these encounters is a primate universal. Modern society taps into a long history of hierarchy formation in which those lower on the scale not only fear the higher-ups but also resent them. We're always ready to wobble the social ladder, a heritage going back to ancestors who roamed the savanna in small egalitarian bands. They gave us asymmetrical reactions to unfairness, always stronger in those who have less than in those who have more. While the latter are not totally indifferent, the ones who get truly worked up, angrily flinging their food away, are invariably the possessors of watery vegetables facing a happy few who gorge themselves on sugary fruits.

Robin Hood had it right. Humanity's deepest wish is to spread the wealth.

Crooked Timber

Out of the crooked timber of humanity no straight thing was ever made.
—IMMANUEL KANT, 1784

We have always known that heedless self-interest was bad morals; we know now that it is bad economics.
—FRANKLIN D. ROOSEVELT, 1937

Asked by a religious magazine what I would change about the human species "if I were God," I had to think hard. Every biologist knows the law of unintended consequences, a close cousin of Murphy's law. Any time we fiddle with an ecosystem by introducing new species, we create a mess. Whether it is the introduction of the Nile perch to Lake Victoria, the rabbit to Australia, or kudzu to the southeastern United States, I am not sure we've ever brought improvement.

Each organism, including our own species, is a complex system in and of itself, so why would it be any easier to avoid unintended consequences? In his utopian novel *Walden Two,* B. F. Skinner thought humans could achieve greater happiness and productivity if parents stopped spending extra time with their children and people refrained from thanking one another. They were allowed to feel indebted to their community, but not to one another. Skinner proposed other pe-culiar codes of conduct, but those two specifically struck me as blows to the pillars of any society: family ties and reciprocity. Skinner must have thought he could improve upon human nature. Along similar lines, I once heard a psychologist seriously propose that we should train children to hug one another several times a day, because isn't hug-ging by all accounts a positive behavior that fosters good relations? It is, but who says that hugs performed on command work the same? Don't we risk turning a perfectly meaningful gesture into one that we can't trust anymore?

We have seen in Romanian orphanages what happens when chil-dren are subjected to the baby-factory ideas of behaviorist psychology. I remain deeply suspicious of any "restructuring" of human nature even though the idea has enjoyed great appeal over the ages. In 1922, Leon Trotsky described the prospect of a glorious New Man:

> There is no doubt whatever but that the man of the future, the citi-zen of the commune, will be an exceedingly interesting and attrac-tive creature, and that his psychology will be very different from ours.

Marxism foundered on the illusion of a culturally engineered human. It assumed that we are born as a tabula rasa, a blank slate, to be filled in by conditioning, education, brainwashing, or whatever we call it, so that we're ready to build a wonderfully cooperative society. A similar illusion plagued the U.S. feminist movement, which (unlike its *vive-la-différence* European counterpart) assumed that gender roles were ready for a complete overhaul. At around the same time, a fa-mous sexologist proposed that a boy who'd lost part of his penis be surgically castrated and raised as a girl, and predicted that he'd be perfectly happy. This "experiment" produced a deeply confused indi-vidual, who committed suicide years later. One can't just ignore the biology of gender identity. In the same way, our species has behav-ioral tendencies that no culture has ever been able to do away with.

As noted by Immanuel Kant, human nature is no more amenable to carving and shaping than is the toughest tree root.

Have you ever noticed how the worst part of someone's personality is often also the best? You may know an anally retentive, detail-oriented accountant who never cracks a joke, nor understands any, but this is in fact what makes him the perfect accountant. Or you may have a flamboyant aunt who constantly embarrasses everyone with her big mouth, yet is the life of every party. The same duality applies to our species. We certainly don't like our aggressiveness—at least on most days—but would it be such a great idea to create a society without it? Wouldn't we all be as meek as lambs? Our sports teams wouldn't care about winning or losing, entrepreneurs would be impossible to find, and pop stars would sing only boring lullabies. I'm not saying that aggressiveness is good, but it enters into everything we do, not just murder and mayhem. Removing human aggression is thus something to consider with care.

Humans are bipolar apes. We have something of the gentle, sexy bonobo, which we may like to emulate, but not too much; otherwise the world might turn into one giant hippie fest of flower power and free love. Happy we might be, but productive perhaps not. And our species also has something of the brutal, domineering chimpanzee, a side we may wish to suppress, but not completely, because how else would we conquer new frontiers and defend our borders? One could argue that there would be no problem if *all* of humanity turned peaceful at the same time, but no population is stable unless it's immune to invasions by mutants. I'd still worry about that one lunatic who gathers an army and exploits the soft spots of the rest.

So, strange as it may sound, I'd be reluctant to radically change the human condition. But if I could change one thing, it would be to expand the range of fellow feeling. The greatest problem today, with so many different groups rubbing shoulders on a crowded planet, is excessive loyalty to one's own nation, group, or religion. Humans are capable of deep disdain for anyone who looks different or thinks another way, even between neighboring groups with almost identical

DNA, such as the Israelis and Palestinians. Nations think they are superior to their neighbors, and religions think they own the truth. When push comes to shove, they are ready to thwart or even eliminate one another. In recent years, we have seen two huge office towers brought down by airplanes deliberately flown into them as well as massive bombing raids on the capital of a nation, and on both occasions the deaths of thousands of innocents was celebrated as a triumph of good over evil. The lives of strangers are often considered worthless. Asked why he never talked about the number of civilians killed in the Iraq War, U.S. defense secretary Donald Rumsfeld answered: "Well, we don't do body counts on other people."

Empathy for "other people" is the one commodity the world is lacking more than oil. It would be great if we could create at least a modicum of it. How this might change things was hinted at when, in 2004, Israeli justice minister Yosef Lapid was touched by images of a Palestinian woman on the evening news. "When I saw a picture on the TV of an old woman on all fours in the ruins of her home looking under some floor tiles for her medicines, I did think, 'What would I say if it were my grandmother?'" Even though Lapid's sentiments infuriated the nation's hard-liners, the incident showed what happens when empathy expands. In a brief moment of humanity, the minister had drawn Palestinians into his circle of concern.

If I were God, I'd work on the reach of empathy.

Russian Doll

Fostering empathy isn't made easier by the entrenched opinion in law schools, business schools, and political corridors that we are essentially competitive animals. Social Darwinism may be dismissed as old hat, a leftover of the Victorian era, but it's still very much with us. A 2007 column by David Brooks in The New York Times ridiculed social government programs: "From the content of our genes, the nature of our neurons and the lessons of evolutionary biology, it has become clear that nature is filled with competition and conflicts of interest." Conservatives love to think this.

I am not saying that their viewpoint lacks any substance, but anyone looking for a rationale of how to structure society should realize that this is only half the truth. It misses by a mile the intensely social nature of our species. Empathy is part of our evolution, and not just a recent part, but an innate, age-old capacity. Relying on automated sensitivities to faces, bodies, and voices, humans empathize from day one. It's really not as complex a skill as it has been made out to be, such as when empathy is said to rest on the attribution of mental states to others, or the ability to consciously recall one's own experiences. No one denies the importance of these higher strata of empathy, which develop with age, but to focus on them is like staring at a splendid cathedral while forgetting that it's made of bricks and mortar.

Martin Hoffman, who has written extensively on this topic, rightly noted that our relations with others are more basic than we think: "Humans must be equipped biologically to function effectively in many social situations without undue reliance on cognitive processes." Even though we are certainly capable of imaginative ways of getting into someone else's head, this is not how we operate most of the time. When we pull a crying child onto our lap, or exchange an understanding smile with a spouse, we're engaged in everyday empathy that is rooted as much in our bodies as in our minds.

In my attempt to strip empathy down to its bare bones, I've made nonhumans an explicit part of the discussion. Not everyone agrees. Some scientists turn into "hear no evil, speak no evil" monkeys, slamming their hands over their ears and mouths, as soon as talk turns to the internal states of other animals. Putting emotional labels on human behavior is fine, but when it comes to animals, we're supposed to suppress this habit. The reason most of us find this almost impossible to do is that humans "mentalize" automatically. Mentalization offers a shortcut to behavior around us. Instead of making piecemeal observations of the way our boss reacts to our late arrival (he frowns, gets red in the face, bangs the table, and so on), we integrate all of this information into a single evaluation (he is mad). We frame the behavior around us according to perceived goals, desires, needs, and emotions. This works great with our boss (even though it hardly improves our situation), and applies equally well to a dog who bounds toward us with wagging tail versus another dog who growls at us with lowered head and bristling fur. We call the first dog "happy" and the second "angry," even though many scientists scoff at the implication of mental states. They prefer terms such as "playful" or "aggressive." The poor dogs are doing everything to make their feelings known, yet science throws itself into linguistic knots to avoid mentioning them.

Obviously, I don't agree with this caution. For the Darwinist, there is nothing more logical than the assumption of emotional continuity. Ultimately, I believe that the reluctance to talk about animal emotions has less to do with science than religion. And not just any religion, but particularly religions that arose in isolation from animals that look like us. With monkeys and apes around every corner, no rain forest culture has ever produced a religion that places humans outside of nature. Similarly, in the East—surrounded by native primates in India, China, and Japan—religions don't draw a sharp line between humans and other animals. Reincarnation occurs in many shapes and forms: A man may become a fish and a fish may become God. Monkey gods, such as Hanuman, are common. Only the Judeo-Christian religions place humans on a pedestal, making them the only species with a soul. It's not hard to see how desert nomads might have arrived at this view. Without animals to hold up a mirror to them, the notion that we're alone came naturally to them. They saw themselves as created in God's image and as the only intelligent life on earth. Even today, we're so convinced of this that we search for other such life by training powerful telescopes on distant galaxies.

It's extremely telling how Westerners reacted when they finally did get to see animals capable of challenging these notions. When the first live apes went on display, people couldn't believe their eyes. In 1835, a male chimpanzee arrived at London Zoo, clothed in a sailor's suit. He was followed by a female orangutan, who was put in a dress. Queen Victoria went to see the exhibit, and was appalled. She called the apes "frightful, and painfully and disagreeably human." This was a widespread sentiment, and even nowadays I occasionally meet people who call apes "disgusting." How can they feel like this unless apes are telling them something about themselves that they don't want to hear? When the same apes at the London Zoo were studied by the young Charles Darwin, he shared the queen's conclusion but without her revulsion. Darwin felt that anyone convinced of man's superiority ought to go take a look at these apes.

All of this occurred in the not too distant past, long after Western religion had spread its creed of human exceptionalism to all corners of knowledge. Philosophy inherited the creed when it blended with theology, and the social sciences inherited it when they emerged out of philosophy. After all, psychology was named after Psykhe, the Greek goddess of the soul. These religious roots are reflected in continued resistance to the second message of evolutionary theory. The first is that all plants and animals, including ourselves, are the product of a single process. This is now widely accepted, also outside biology. But the second message is that we are continuous with all other life forms, not only in body but also in mind. This remains hard to swallow. Even those who recognize humans as a product of evolution keep searching for that one divine spark, that one "huge anomaly" that sets us apart. The religious connection has long been pushed to the subconscious, yet science keeps looking for something special that we as a species can be proud of.

When it comes to characteristics that we *don't* like about ourselves, continuity is rarely an issue. As soon as people kill, abandon, rape, or otherwise mistreat one another we are quick to blame it on our genes. Warfare and aggression are widely recognized as biological traits, and no one thinks twice about pointing at ants or chimps for parallels. It's only with regard to noble characteristics that continuity is an issue, and empathy is a case in point. Toward the end of a long career, many a scientist cannot resist producing a synopsis of what distinguishes us from the brutes. American psychologist David Premack focused on causal reasoning, culture, and the taking of another's perspective, while his colleague Jerome Kagan mentioned language, morality, and yes, empathy. Kagan included consolation behavior, such as a child embracing his mother, who has hurt herself. This is indeed a great example, but of course hardly restricted to our species. My main point, however, is not whether the proposed distinctions are real or imagined, but why all of them need to be in our favor. Aren't humans at least equally special with respect to torture, genocide, deception, exploitation, indoctrination, and environmental destruction? Why does every list of human distinctiveness need to have the flavor of a feel-good note?

There is a deeper problem, though, which brings me back to the status we assign empathy in society. If being sensitive to others were truly limited to our species, this would make it a young trait, something we evolved only recently. The problem with young traits, however, is that they tend to be experimental. Consider the human back. When our ancestors started walking on two legs, their backs straightened and assumed a vertical position. In doing so, backs became the bearers of extra weight. Since this is not what the vertebral column was originally designed for, chronic back pain became our species' universal curse.

If empathy were truly like a toupee put on our head yesterday, my greatest fear would be that it might blow off tomorrow. Linking empathy to our frontal lobes, which achieved their extraordinary size only in the last couple of million years, denies how much it is a part of who and what we are. Obviously, I believe the exact opposite, which is that empathy is part of a heritage as ancient as the mammalian line. Empathy engages brain areas that are more than a hundred million years old. The capacity arose long ago with motor mimicry and emotional contagion, after which evolution added layer after layer, until our ancestors not only felt what others felt, but understood what others might want or need. The full capacity seems put together like a Russian doll. At its core is an automated process shared with a multi-

Perspective-Taking
targeted helping

Concern for Others
consolation

State-Matching
emotional contagion

Empathy is multilayered, like a Russian doll, with at its core the ancient tendency to match another's emotional state. Around this core, evolution has built ever more sophisticated capacities, such as feeling concern for others and adopting their viewpoint.

tude of species, surrounded by outer layers that fine-tune its aim and reach. Not all species possess all layers: Only a few take another's perspective, something we are masters at. But even the most sophisticated layers of the doll normally remain firmly tied to its primal core.

Evolution rarely throws out anything. Structures are transformed, modified, co-opted for other functions, or tweaked in another direction. Thus the frontal fins of fish became the front limbs of land animals, which over time turned into hooves, paws, wings, and hands. They also became the flippers of mammals that returned to the water. This is why to the biologist a Russian doll is such a satisfying plaything, especially if it has a historical dimension. I own a wooden doll of former Russian president Vladimir Putin. He is depicted on the outside, and within him we discover, in this order, Yeltsin, Gorbachev, Brezhnev, Khrushchev, Stalin, and Lenin. Finding a tiny little Lenin and Stalin within Putin will hardly surprise most political analysts. But the same is true for biological traits: The old remains present in the new. This is relevant to the story of empathy since it means that even our most thoughtful reactions to others share core processes with the reactions of young children, other primates, elephants, dogs, and rodents.

I derive great optimism from empathy's evolutionary antiquity. It makes it a robust trait that will develop in virtually every human being so that society can count on it and try to foster and grow it. It is a human universal. In this regard, it's like our tendency to form social hierarchies, which we share with so many animals, and which we don't need to teach or explain to children: They arrange themselves spontaneously into pecking orders before we know it. What society does instead is either enhance this tendency, as is done in male bastions such as the church or the military, or counter it, as done in small-scale egalitarian societies. In the same way, human empathy is so ingrained that it will almost always find expression, giving us material to work with either by countering it, as we do when we dehumanize our enemies, or by enhancing it, as when we urge a child who is hogging all the toys to be more considerate of her playmates.

We may not be able to create a New Man, but we're remarkably good at modifying the old one.

The Dark Side

Have you ever heard of an organization that appeals to empathy in order to fight the lack of it? That the world needs such an organization, known as Amnesty International, says a lot about the dark side of our species. British author J. K. Rowling describes an experience during her time working at Amnesty International's headquarters in London:

> As long as I live I shall remember walking along an empty corridor and suddenly hearing, from behind a closed door, a scream of pain and horror such as I have never heard since. The door opened, and a researcher poked out her head and told me to run and make a hot drink for the young man sitting with her. She had just given him the news that in retaliation for his own outspokenness against his country's regime, his mother had been seized and executed.

If empathy were purely intellectual, a product of our prefrontal cortex, the *Harry Potter* author wouldn't have felt anything special at hearing the man's scream, nor would she have remembered it all of her life. But empathy goes a thousand times deeper: It touches parts of the brain where screams don't just register, but induce fear and loathing. We literally *feel* a scream. We should be grateful for this, because otherwise there would be no reason for empathy to be used for good. In and of itself, taking another's perspective is a neutral capacity: It can serve both constructive and destructive ends. Crimes against humanity often rely on precisely this capacity.

Torture requires an appreciation of what others think or feel. To attach electrodes to the genitals of prisoners, hang them upside down for prolonged periods of time, simulate drowning during so-called "waterboarding," or urinate on their Bible or Koran rests on our ability to assume their viewpoint and realize what will hurt or aggravate them the most. Go to any medieval torture museum, with its garrotes, chairs with spikes, head crushers, and thumb screws, and see what human imagination has wrought in the service of suffering. Our species even engages in vicarious torture. To rape a woman in front of her husband is not only brutal to her, but also a way of tormenting him. It exploits the bond one person feels with another. Cruelty, too, rests on perspective-taking.

One mental illness is marked by a permanent disconnect between perspective-taking and the deeper regions of empathy. The label of "psychopath" is often associated with violence, such as serial killers Ted Bundy and Harold Shipman, or mass murderers, such as Joseph Stalin, Benito Mussolini, and Saddam Hussein. But psychopathy takes many forms. The condition is defined by an antisocial attitude without loyalty to anyone except the self. Think of the boyfriend who leaves a woman after having emptied out her bank account, but who returns months later with a bouquet of roses for a tearful reunion so that he can move back in and start the whole cycle over again. Or the CEO who makes loads of money over the backs of others, even talking trusting employees into holding on to company stock at the very moment that he himself is unloading his shares, as Kenneth Lay did before Enron's collapse in 2001. People without mercy and morals are all around us, often in prominent positions. These snakes in suits, as one book title labels them, may represent a small percentage of the population, but they thrive in an economic system that rewards ruthlessness.

The comparison with snakes is apt, since psychopaths seem to lack the Russian doll's old mammalian core. They do possess all of its cognitive outer layers, allowing them to understand what others want and need as well as what their weaknesses are, but they couldn't care less about how their behavior impacts them. According to one theory, they suffer from a developmental disorder that puts them on a wrong learning track early in life. If a normal child makes a younger sibling cry, it will be troubled by the other's distress. The result is aversive conditioning: Children learn not to pester or hit others. Like all social animals, they discover that if you want to have fun, it isn't a good idea to make your playmate yell out in pain. With age, children become gentler with younger, weaker parties, controlling their strength the way a large dog plays with a smaller one or a cat, or for that matter, how a 1,200-pound polar bear plays with a husky. The young psychopath, by contrast, starts life without this sensitivity. Nothing in an encounter with a vulnerable party, least of all teary protest, tells him to back off. On the contrary, all that he ever seems to learn is that hurting others yields benefits. Isn't it an excellent way of getting toys or winning games? The young psychopath only sees the upside of defeating others. The result is a different learning curve, one that culminates in manipulation and intimidation without the slightest worry about the pain it may cause.

Being gentle with vulnerable others is something all children and animals learn during play, such as a polar bear here with a sled dog.

A lot of trouble in the world can be traced to people whose Russian doll is an empty shell. Like aliens from another planet, they are intellectually capable of adopting another's viewpoint without any of the accompanying feelings. They successfully fake empathy. Whenever they achieve power, which they often do thanks to their Machiavellian skills, their disdain for truth and morality allows them to manipulate others into carrying out their evil plans. Their authority overrides the better judgment of underlings so that sometimes an entire nation, for example Germany in the previous century, falls for the cruel fantasies of a charismatic psychopath.

What makes it so hard to wrap our heads around this mental disorder is that we can't imagine being immune to the suffering of others. Mark Rowlands, in *The Philosopher and the Wolf*, describes how hard he found it to treat his house pet. Brenin, the wolf, needed regular cleaning and antibiotics for an infection near his anus, which was excruciatingly painful for him, and by extension for his master. This is what empathy does: It makes it tough to hurt others even if done for a good reason. In a philosophical twist, Rowlands reflected on Tertullian of Carthage, a theologian of early Christianity. This zealous defender of the truth had a most unusual description of heaven. While hell was a place of torture, heaven was a balcony from which the saved ones could watch hell, thus enjoying the spectacle of others frying. One must indeed be close to psychopathy to imagine eternal spite as a blessed state. For most of us, it's almost harder to watch the suffering of others than to suffer ourselves, which is why Rowland adds, "In those days that Brenin was dying, I used to think that this was what hell was—being forced to torture a wolf I loved."

All of this is to contrast those who do empathize with those who don't. This is not to say, though, that those who do, do so all the time. What kind of life would we have if we shared in every form of suffering in the world? Empathy needs both a filter that makes us select what we react to, and a turn-off switch. Like every emotional reaction, it has a "portal," a situation that typically triggers it or that we allow to trigger it. Empathy's chief portal is identification. We're ready to share the feelings of someone we identify with, which is why we do so easily with those who belong to our inner circle: For them the portal is always ajar. Outside this circle, things are optional. It depends on whether we can afford being affected, or whether we want to be. If we notice a beggar in the street, we can choose to look at him, which may arouse our pity, or we can look away, even walk to the other side of the street, to avoid facing him. We have all sorts of ways to open or close the portal.

The moment we buy a movie ticket, we choose to identify with the leading character, thus making ourselves vulnerable to empathizing. We swoon when she falls in love or leave the theater in tears because of her untimely death, even though it's just a character played by someone we don't personally know. On the other hand, we sometimes deliberately shut the portal, such as when we suppress identification with a declared enemy group. We do so by removing their individuality, defining them as an anonymous mass of unpleasant, inferior specimens of a different taxonomic group. Why should we put up with those dirty "cockroaches" (the Hutus about the Tutsis) or disease-ridden "rats" (the Nazis about the Jews)? Called the fifth horseman of the apocalypse, dehumanization has a long history of excusing atrocities.

Since men are the more territorial gender, and overall more confrontational and violent than women, one would expect them to have the more effective turn-off switch. They clearly do have empathy, but perhaps apply it more selectively. Cross-cultural studies confirm that women everywhere are considered more empathic than men, so much so that the claim has been made that the female (but not the male) brain is hardwired for empathy. I doubt that the difference is that absolute, but it's true that at birth girl babies look longer at faces than boy babies, who look longer at suspended mechanical mobiles. Growing up, girls are more prosocial than boys, better readers of emotional expressions, more attuned to voices, more remorseful after having hurt someone, and better at taking another's perspective. When Carolyn Zahn-Waxler measured reactions to distressed family members, she found girls looking more at the other's face, providing more physical comfort, and more often expressing concern, such as asking "Are you okay?" Boys are less attentive to the feelings of others, more action- and object-oriented, rougher in their play, and less inclined to social fantasy games. They prefer collective action, such as building something together.

Men can be quite dismissive of empathy. It's not particularly manly to admit to it, and one reason why it has taken so long for research in this area to take off is undoubtedly that academics saw empathy as a bleeding-heart topic associated with the weaker sex. That this is a traditional attitude is exemplified by Bernard de Mandeville, a Dutch philosopher and satirist of the eighteenth century who saw "pity" as a character flaw:

> Pity, though it is the most gentle and the least mischievous of all our passions, is yet as much a frailty of our nature, as anger, pride, or fear. The weakest minds have generally the greatest share of it, for which reason none are more compassionate than women and children. It must be owned, that of all our weaknesses it is the most amiable, and bears the greatest resemblance to virtue; nay, without a considerable mixture of it, the society could hardly subsist.

The tortuous nature of this statement is understandable for a cynic who gave the world its first greed creed. Mandeville didn't know where to fit the tender emotions, but was at least honest enough to recognize that society would be in trouble without them.

Despite the association of empathy with women rather than men, some studies paint a more complex picture. They call gender differences in this regard "exaggerated," even "nonexistent." These claims are puzzling given the well-documented difference between boys and girls. Are we to believe that the sexes converge with age? My guess is that they don't, and that the confusion stems from the way men and women have been tested by psychologists. Asked about loved ones, such as their parents, wife, children, and close friends, most men are plenty empathic. The same applies in relation to unfamiliar, neutral parties. Men are perfectly willing to empathize under such circumstances, the way they often can't keep their eyes dry in romantic or tragic movies. With their portal open, men can be just as empathic as women.

But things change radically when men enter a competitive mode, such as when they're advancing their interests or career. Suddenly, there's little room for softer feelings. Men can be brutal toward potential rivals: Anyone who stands in the way has to be taken down. Sometimes the physicality slips out, such as when Jesse Jackson, the longtime African American alpha male, expressed his feelings about the new kid on the block, Barack Obama, in 2008. In surreptitiously taped comments on a television show, Jackson said about Obama that he'd like to "cut his nuts off." At other times, things literally get physical, such as the way the head of Microsoft, Steve Ballmer, reacted to hearing that a senior engineer of his company was going to work for his competitors at Google. Ballmer was said to have picked up a chair and thrown it forcefully across the room, hitting a table. After this chimpanzee-like display, he launched into a tirade about how he was going to fucking kill those Google boys.

Many men love action movies, which would be a disastrous experience if they had any sympathy for their hero's adversaries. The villains are blown to pieces, riddled with bullets, thrown into shark-infested pools, and pushed out of flying airplanes. None of this bothers the audience. On the contrary, they pay to watch the carnage. Sometimes the hero himself gets caught, then is strung up in chains and tormented with burning coals, which makes the audience squirm. But since it's just a movie, he always gets out and exacts revenge, which is sweeter the nastier it gets.

Male primates may be similar. Robert Sapolsky, who occasionally tranquilizes wild baboons, learned the hard way how dangerous it is to dart a male in front of his rivals. As soon as the darted male's walk becomes unsteady, others close in, seeing a perfect opportunity to get him. There is no problem with the females, but male baboons are always ready to take advantage of another's weakness. This is why vulnerabilities are hidden. I have known male chimps who went into unusually vigorous intimidation displays during times that they were sick or injured. They'd be licking their wounds one minute, looking miserable, but then their main rival would show up and suddenly they'd be full of muscle power, at least for the few minutes that mattered. In the same way, I imagine a group of human ancestors in which men camouflaged for as long as possible any limp, reduced eyesight, or loss of stamina so as not to give the others any ideas. This is why the Kremlin used to prop up its ailing leaders, and why enterprises sometimes hesitate to disclose the health problems of their CEOs, as Apple did with Steve Jobs. In modern society, it's often said that men don't go to the doctor as easily as women because they have been socialized to act tough, but what if there's a much deeper reason? Perhaps males always feel surrounded by others hoping for them to stumble.

The opposite occurs when men are in the company of trusted parties. Often this will be a wife or girlfriend, but it extends to their best male buddies. Men value nothing more than loyalty, and in these situations they do show vulnerability, which elicits sympathy. There's plenty of this among men on the same team, such as in sports or the military, and I once saw an interesting sign of it among chimpanzees. An old male had partnered with a younger one, who was more muscular and energetic. The old male had helped his friend reach the top, but one day this new leader nevertheless bit his partner in a conflict over a female. This was not very smart, because his position depended on the old male's support. Naturally, the young male did a lot of grooming to smooth things over, but the old fox—perhaps the most cunning chimp I've ever known—couldn't resist rubbing in how much he had been hurt. For days, he limped pitifully each time he was in view of the young leader, whereas away from him he walked normally. Now, why put up an act like this, if sympathy plays no role in male relations?

It's possible, then, that male sensitivity to others is conditional, aroused mostly by family and friends. For those who don't belong to the inner circle, and especially those who act like rivals, the portal remains closed and the empathy switch turned off. Neuroscience supports this idea for humans. A German investigator, Tania Singer, tested men and women in a brain scanner while they could see another in pain. Both sexes commiserated with the other: The pain areas in their own brains lit up when they saw the other's hand getting mildly shocked. It was as if they felt the sting themselves. But this happened only if the partner was someone likable and with whom they had played a friendly game. Things changed drastically if the partner had played unfairly in the previous game. Now the subjects felt cheated, and seeing the other in pain had less of an effect. Women still showed some empathy, but the men had nothing left. On the contrary, if men saw the unfair player getting shocked, their brain's pleasure centers lit up. They had moved from empathy to justice, and seemed to enjoy the other's punishment. Perhaps there exists a Tertullian heaven after all, at least for men, where they watch their enemies roast in flames.

Nevertheless, men seem unable to turn their empathy switch completely down to zero. One of the most illuminating books I have read in recent years is *On Killing*, by Lieutenant Colonel Dave Grossman, who served in the U.S. Army. Grossman follows in the footsteps of Leo Tolstoy, who gave us *War and Peace* and said that he was more interested in how and why soldiers kill, and what they feel while doing so, than in how generals arrange their armies on the battlefield. To actually kill someone is, of course, quite different from watching a movie about it, and in this regard the data tell us something few would have suspected: Most men lack a killer instinct.

It is a curious fact that the majority of soldiers, although well armed, never kill. During World War II, only one out of every five U.S. soldiers actually fired at the enemy. The other four were plenty courageous, braving grave danger, landing on the beaches, rescuing comrades under fire, fetching ammunition for others, and so on, yet they failed to fire their weapons. One officer reported that "squad leaders and platoon sergeants had to move up and down the firing line kicking men to get them to fire. We felt like we were doing good to get two or three men out of a squad to fire." Similarly, it has been calculated that during the Vietnam War, U.S. soldiers fired more than fifty thousand bullets for every enemy soldier killed. Most bullets must have been fired into the air.

This recalls the famous Stanley Milgram experiment in which

human subjects were asked to deliver high-voltage shocks to others. They obeyed the experimenter to a surprising degree, but began to cheat as soon as he was called away. Subjects still acted as if they were giving shocks, but were now feigning punishment by administering much milder ones. Grossman himself draws a comparison with New Guinean tribes where the men are excellent shots with bow and arrow during the hunt, but when they go to war they remove the feathers from their arrows, thus rendering them useless. They prefer to fight with inaccurate weapons, knowing that their enemies will have taken the same step.

Killing or hurting others is something we find so horrendous that wars are often a collective conspiracy to miss, an artifice of incompetence, a game of posturing rather than an actual hostile confrontation. Nowadays, this is not always realized, given that wars can be fought at a distance almost like a computer game, which eliminates most of these natural inhibitions. But actual killing at close range has no glory, no pleasure, and is something the typical soldier tries to avoid at all cost. Only a small percentage of men—perhaps 1 or 2 percent—does the vast majority of killing during a war. This may be the same category of humans discussed before, the one immune to the suffering of others. Most soldiers report a deep revulsion: They vomit at the sight of dead enemies, and end up with haunting memories. Lifelong combat trauma was already known to the ancient Greeks, as reflected in Sophocles' plays about the "divine madness" that we now call post-traumatic stress disorder. Decades after a war, veterans still can't hold back tears when asked about the killings they have witnessed. The sorrow and repulsion associated with these images is triggered by our species' natural body language, similar to the scream that Rowling couldn't get out of her head. This is also what makes it so hard to apply lethal force at close range: "The average soldier has an intense resistance to bayoneting his fellow man, and this act is surpassed only by the resistance to *being* bayoneted."

So, anyone who would like to use war atrocities as an argument against human empathy needs to think twice. The two aren't mutually exclusive, and it's important to consider how hard most men find it to pull the trigger. Why would this be, if not for empathy with their fellow human beings? Warfare is psychologically complex, and seems more a product of hierarchy and following orders than of aggression and lack of mercy. We are definitely capable of it, and do kill for our country, but the activity conflicts at the deepest level with our humanity. Even "scorched earth" Union General William Sherman had nothing good to say about it:

> I am sick and tired of war. Its glory is all moonshine. It is only those who have neither fired a shot nor heard the shrieks and groans of the wounded who cry aloud for blood, for vengeance, for desolation. War is hell.

The Invisible Helping Hand

One of the first debates about the role of empathy in human life reached us more than two millennia ago from a Chinese sage, Mencius, a follower of Confucius. Mencius saw empathy as part of human nature, famously stating that everyone is born with a mind that cannot bear to see the suffering of others.

In one of Mencius's stories, the king watches an ox being led past his palace. The king wants to know what's going on, and is told that the ox is on its way to being slaughtered so that its blood may be used for a ceremony. The king can't stand the ox's frightened appearance, however, which to him suggests that it realizes what is about to happen. He orders that the ox be saved. But not wanting to cancel the ceremony, he proposes to sacrifice a sheep instead.

Mencius is unimpressed by the king's pity for the ox, telling him that he seems as much concerned with his own tender feelings as the animal's fate:

> You saw the ox, and had not seen the sheep. So is the superior man affected towards animals, that, having seen them alive, he cannot

bear to see them die; having heard their dying cries, he cannot bear to eat their flesh. Therefore he keeps away from his slaughter-house and cook-room.

We care more about what we see firsthand than about what remains out of sight. We're certainly capable of feeling for others based on hearing, reading, or thinking about them, but concern based purely on the imagination lacks strength and urgency. Hearing the news that a good friend has fallen ill and is suffering in a hospital, we'll sympathize. But our worries intensify tenfold when we actually stand at his bedside and notice how pale he looks, or how much trouble he has breathing.

Mencius made us reflect on the origin of empathy, and how much it owes to bodily connections. These connections also explain the trouble we have empathizing with outsiders. Empathy builds on proximity, similarity, and familiarity, which is entirely logical given that it evolved to promote in-group cooperation. Combined with our interest in social harmony, which requires a fair distribution of resources, empathy put the human species on a path toward small-scale societies that stress equality and solidarity. Nowadays, most of us live in much larger societies, where this emphasis is harder to maintain, but we still have a psychology that feels most comfortable with these outcomes.

A society based purely on selfish motives and market forces may produce wealth, yet it can't produce the unity and mutual trust that make life worthwhile. This is why surveys measure the greatest happiness not in the wealthiest nations but rather in those with the highest levels of trust among citizens. Conversely, the trust-starved climate of modern business spells trouble and has recently made many people deeply unhappy by wiping out their savings. In 2008, the world's financial system collapsed under the weight of predatory lending, reporting of nonexisting profits, pyramid schemes, and reckless betting with other people's money. One of the system's architects, former Federal Reserve chairman Alan Greenspan, said that he had no idea

this might happen. In response to a grilling by a U.S. House committee, he acknowledged that his vision had been flawed: "That is precisely the reason I was shocked because I'd been going for forty years
or more with very considerable evidence that it was working exceptionally well."

The mistake of Greenspan and other supply-side economists was
to assume that, even though the free market by itself is no moral enterprise, it would steer society toward a state in which everyone's interests were optimally served. Hadn't their demigod, Milton Friedman,
declared that social responsibility conflicts with freedom? And hadn't
an even higher authority, Adam Smith, given them the metaphor of
the "invisible hand," according to which even the most selfish motives
will automatically advance the greater good? The free market knows
what is best for us. The baker needs income, his clients need bread,
and voilà, both parties stand to gain from their transaction. Morality
has nothing to do with it.

Unfortunately, these references to Smith are selective. They leave
out an essential part of his thinking, which is far more congenial to
the position I have taken throughout this book; namely, that reliance
on greed as the driving force of society is bound to undermine its very
fabric. Smith saw society as a huge machine, the wheels of which are
polished by virtue, whereas vice causes them to grate. The machine
just won't run smoothly without a strong community sense in every
citizen. Smith frequently mentioned honesty, morality, sympathy,
and justice, seeing them as essential companions to the invisible hand
of the market.

In effect, society depends on a second invisible hand, one that
reaches out to others. The feeling that one human being cannot be indifferent to another if we wish to build a community true to the
meaning of the word is the other force that underlies our dealings
with one another. The evolutionary antiquity of this force makes it all
the more surprising just how often it is being ignored. Do business
schools teach ethics and obligations to the community in any context
other than how it may advance business? Do they pay equal attention

to stakeholders and shareholders? And why does the "dismal science" attract so few female students, and has never produced a female Nobelist? Could it be that women feel no connection to the caricature of a rational being whose only goal in life is to maximize profit? Where are human relations in all of this?

It's not as if we're asking our species to do anything foreign to it by building on the old herd instinct that has kept animal societies together for millions of years. And here I don't mean that we should blindly follow one another, but that we have to stick together: We can't just scatter in all directions. Every individual is connected to something larger than itself. Those who like to depict this connection as contrived, as not part of human biology, don't have the latest behavioral and neurological data on their side. The connection is deeply felt and, as Mandeville had to admit, no society can do without it.

First of all, there are the occasions where others need aid and we have a chance to offer it in the form of food banks, disaster relief, elderly care, summer camps for poor children, and so on. Measured by volunteer community services, Western societies seem to be in great shape indeed, and have plenty of compassion to go around. But the second area where solidarity counts is the common good, which includes health care, education, infrastructure, transportation, national defense, protection against nature, and so on. Here the role of empathy is more indirect, because no one would want to see such vital pillars of society depend purely on the warm glow of kindness.

The firmest support for the common good comes from enlightened self-interest: the realization that we're all better off if we work together. If we don't benefit from our contributions now, then at least potentially we will in the future, and if not personally, then at least via improved conditions around us. Since empathy binds individuals together and gives each a stake in the welfare of others, it bridges the world of direct "what's in it for me?" benefits and collective benefits, which take a bit more reflection to grasp. Empathy has the power to open our eyes to the latter by attaching emotional value to them. Let me give two concrete examples.

When Hurricane Katrina hit Louisiana in 2005, our television screens showed massive human despair. The disaster was exacerbated by the gross incompetence of agencies that were supposed to deal with its aftermath and by the cold detachment of politicians at the highest levels. The rest of the nation watched with a mixture of horror, pity, and worry. The worry was not without self-interest, because obviously the way one mammoth disaster is being handled tells us something about how others may be handled in the future, including ones that hit us. The lackluster official response had a twofold impact: amazing generosity from the public, and a shift in perception about governmental responsibility. Until Katrina, the nation's leadership had gotten away with its everyone-for-himself philosophy, but the catastrophe raised serious doubts about it. As Barack Obama said three years later, "We are more compassionate than a government that lets veterans sleep on our streets and families slide into poverty; that sits on its hands while a major American city drowns before our eyes."

Another example of how empathy figures into public policy debates concerns abolitionism. Again, the impetus came not just from imagining how bad slavery was, but from firsthand observation of its cruelty. Abraham Lincoln was plagued by negative feelings, as he explained in a letter to a slave-owning friend in Kentucky:

> In 1841 you and I had together a tedious low-water trip, on a Steam Boat from Louisville to St. Louis. You may remember, as I well do, that from Louisville to the mouth of the Ohio, there were, on board, ten or a dozen slaves, shackled together with irons. That sight was a continued torment to me; and I see something like it every time I touch the Ohio, or any other slave-border. [It is] a thing which has, and continually exercises, the power of making me miserable.

Such sentiments were of course not limited to Lincoln and motivated many others to fight slavery. One of the most potent weapons of the abolitionist movement were drawings of slave ships and their human cargo, which were disseminated to generate empathy and moral outrage. The role of compassion in society is therefore not just one of sacrificing time and money to relieve the plight of others, but also of pushing a political agenda that recognizes everyone's dignity. Such an agenda helps not merely those who need it most, but also the larger whole. One can't expect high levels of trust in a society with huge income disparities, huge insecurities, and a disenfranchised underclass. And remember, trust is what citizens value most in their society.

Obviously, how to achieve this goal cannot be easily inferred from watching animal communities, or even small-scale human societies. The world we live in is infinitely larger and more complex. We need to rely on our well-developed intellect to figure out how to balance individual and collective interests on such a scale. But one instrument that we do have available, and that greatly enriches our thinking, has been selected over the ages, meaning that it has been tested over and over with regard to its survival value. That is our capacity to connect to and understand others and make their situation our own, the way the American people did while watching Katrina victims and Lincoln did when he came eye to eye with shackled slaves.

To call upon this inborn capacity can only be to any society's advantage.

Notes

PREFACE

ix   **Barack Obama:** 2006 Commencement Address at Northwestern University, Northwestern News Service, June 22, 2006.

CHAPTER 1: BIOLOGY, LEFT AND RIGHT

1   **"What is government":** Federalist Paper No. 51 (Rossiter, 1961, p. 322).

2   **"How selfish soever":** Adam Smith, *The Theory of Moral Sentiments* (1759, p. 9).

5   **"failure of citizenship":** Newt Gingrich, address to the Conservative Political Action Conference, March 2, 2007: "How can you have the mess we have in New Orleans, and not have had deep investigations of the federal government, the state government, the city government, and the failure of citizenship in the Ninth Ward, where 22,000 people were so uneducated and so unprepared, they literally couldn't get out of the way of a hurricane"?

8   **"Any animal whatever":** Charles Darwin (1871, pp. 71–72).

9   **It's the bedrock:** An argument worked out in full in *Primates and Philosophers: How Morality Evolved* (de Waal, 2006).

12   **father of behaviorism:** John B. Watson (1878–1958) and B. F. Skinner (1904–90), both interested in animal behavior and its connection to human behavior, founded the influential school of behaviorism.

13   **Watson's crusade:** For Watson's views, which were hardly atypical for

his day, and the figure of Harry Harlow, see Deborah Blum's (2002) illuminating *Love at Goon Park*.

17   **known as the San:** Even though the "Bushmen" label may seem politically incorrect, it is used to denote both Bushmen and Bushwomen. Elizabeth Marshall Thomas (2006, p. 47) explains that this is how this population now calls itself. The label "San" used by anthropologists is apparently a derogatory term derived from the Nama word for "bandits."

18   **stay off the ground:** Less than two million years ago *Homo erectus* still had arboreal adaptations that suggest sleeping in trees for safety (Lordkipanidze et al., 2007).

18   **When they cross a human dirt road:** Kimberley Hockings and coworkers (2006) documented road crossings by wild chimpanzees surrounded by human populations.

21   **reveries of centuries past:** Jean-Jacques Rousseau, who famously said, "Man was born free," explained that "man's first law is to watch over his own preservation; his first care he owes to himself; and as soon as he reaches the age of reason, he becomes the only judge of the best means to preserve himself; he becomes his own master" (*The Social Contract*, 1762, pp. 49–50). Rousseau painted the image of our ancestors asleep under a fruit tree in the jungle: their bellies full, their minds clear of worries. How much of an illusion such a carefree existence is may not have occurred to Rousseau, who bore five children with his live-in maid but dispatched every one of them to an orphanage.

22   **"The story of the human race":** Winston Churchill (1932).

23   **walls of Jericho:** Israeli archeologist Ofer Bar-Yosef (1986) studied the walls of Jericho, noting that the city had no known enemies, that accumulation of debris had made the walls scalable (which should have been prevented if they served military purposes), and that Jericho happens to sit on a sloping plain next to a drainage basin, and hence probably was subject to massive mudflows.

23   **scattered small bands:** A study of mitochondrial DNA suggests near extinction of our species, with total numbers of around two thousand, before the population came back from the brink (Behar et al., 2008). "Tiny bands of early humans developed in isolation from each other for as much as half of our entire history as a species," according to Doron Behar, a genographer in Haifa, Israel (Breitbart.com, April 25, 2008).

23 **stretches of peace and harmony:** Douglas Fry (2006) reviews the anthropological literature on warfare, defined as armed combat between political entities, and challenges the "war assumption" of Winston Churchill and others. Whereas archeological evidence for homicide is abundant (and homicide is common in present-day hunter-gatherers, such as the Bushmen), solid evidence for warfare goes back at most fifteen thousand years. See also John Horgan's "Has Science Found a Way to End All Wars?" (*Discover*, March 2008).

23 **Comparisons with apes:** Bonobos and chimpanzees are our closest primate relatives, with whom we share a common ancestor estimated to have lived 5 or 6 million years ago. Also known as the "make love—not war" primates, bonobos have a reputation for being exceptionally peaceful (de Waal, 1997). Observations of sexual "mingling" at territorial boundaries were first made by Japanese scientists, led by Takayoshi Kano, who devoted his life to fieldwork in the Democratic Republic of the Congo (Kano, 1992). Not a single fatal attack among bonobos, captive or wild, has ever been witnessed, whereas dozens of such attacks have been documented in chimpanzees (e.g., de Waal, 1986; Wrangham and Peterson, 1996). Recent observations of bonobos hunting and killing monkeys have been interpreted as being at odds with this pacific image, but predation is not the same as fighting. Predation is motivated by hunger rather than aggression, and relies on different brain circuits, which explains why herbivores can be quite aggressive. Further see de Waal's "Bonobos, Left & Right" in *eSkeptic* (August 8, 2007).

24 **the average Bushman:** Polly Wiessner (personal communication). For more on how primitive warfare is constrained by ties between communities, see Lars Rodseth and co-workers (1991) and Wiessner (2001).

25 **"It is bad to die":** Elizabeth Marshall Thomas (2006, p. 213).

CHAPTER 2: THE OTHER DARWINISM

27 **"Manchester Newspaper":** Charles Darwin complained in a letter to an eminent geologist about a commentary on Napoleon's exploits in *The Manchester Guardian*, titled "National and Individual Rapacity Vindicated by the Law of Nature" (Letter #2782, May 4, 1860, www.darwinproject.ac.uk).

28 **"believe in evolution":** Debate among Republican presidential candidates at the Ronald Reagan Presidential Library, Simi Valley, California (May 3, 2007).

28   **"the whole effort of nature":** Herbert Spencer (1864, p. 414).

28   **Andrew Carnegie:** Andrew Carnegie (1889) on the law of competition: "While the law may sometimes be hard for the individual, it is best for the race, because it ensures the survival of the fittest in every department."

28   **John D. Rockefeller:** Quoted in Richard Hofstadter's *Social Darwinism in American Thought* (1944, p. 45).

28   **This religious angle:** On the issue of compassion (or lack thereof) in American society, see Candace Clark's *Misery and Company* (1997). About one-third of the U.S. population believes that the rich owe nothing to the poor (Pew Research Center, 2004). The Bible couldn't be clearer, though, urging us to act as "a defense for the helpless, a defense for the needy in his distress, a refuge from the storm, a shade from the heat" (Isaiah 25:4). Fortunately, many religious groups are more faithful to the values of the Bible than to those of Social Darwinism, such as when they run soup kitchens in inner cities or provide massive assistance to victims of natural disasters.

30   **This is why I'm tired:** Not only do Social Darwinists erroneously equate their ideology with evolutionary theory, the opponents of Social Darwinism also don't think twice about blaming evolutionary theory. This confusion remains alive today, as evident from this statement by Ben Stein, an American actor: "Darwinism, perhaps mixed with Imperialism, gave us Social Darwinism, a form of racism so vicious that it countenanced the Holocaust against the Jews and mass murder of many other groups in the name of speeding along the evolutionary process" (www.expelledthemovie.com, October 31, 2007).

31   **novelty seekers:** On Americans as a self-selected people with a certain personality type, see Peter Whybrow's *American Mania* (2005).

32   **"In Europe we habitually":** Alexis de Tocqueville (*Democracy in America*, 1835, p. 284).

32   **Ayn Rand:** In a typical passage of Rand's *Atlas Shrugged* (1957, p. 1059), the novel's main character, John Galt, claims: "Accept the fact that achievement of your happiness is the only moral purpose of your life, and that *happiness* . . . is the proof of your moral integrity, since it is the proof and the result of your loyalty to the achievement of your values."

36   **monkey society was falling apart:** Jessica Flack and co-workers (2005).

37   **"What's wrong with America":** Steve Skvara, sixty years old, was one of millions of Americans who over the past decade lost their health insurance or were bankrupted by health-care costs. He became an instant celebrity with his heartfelt question at the AFL-CIO Presidential Candidates Forum in Chicago (August 7, 2007).

37   **quality of the health care:** The United States pays more for health care per capita than any other nation, yet receives less in return. In terms of overall quality, U.S. health care ranks thirty-seventh in the world (World Health Organization: 2007), and on the most critical health index—average life expectancy—the United States ranks only forty-second (National Center for Health Statistics: 2004). See also Sharon Begley, "The Myth of 'Best in the World'" (Newsweek, March 31, 2008).

38   **Milton Friedman:** From "Capitalism and Freedom" (1962, p. 133).

38   **the Enron's Corporation's sixty-four page:** Michael Miller in Business First of Columbus (March 29, 2002).

39   **Jeff Skilling:** Bethany McLean and Peter Elkind in Smartest Guys in the Room (2003).

39   **Genes can't be any more "selfish":** Philosopher Mary Midgley (1979) acidly compared Dawkins's warnings against his own metaphor to the paternosters of Mafiosi. I myself (de Waal, 1996) have protested overextension of the metaphor, such as in Dawkins's claim: "Let us try to teach generosity and altruism, because we are born selfish," which equates the selfishness of genes with psychological selfishness. It was good to see this particular sentence retracted in the 30th Anniversary Edition of The Selfish Gene (2006, p. ix).

40   **the epic drought in Georgia:** Governor Sonny Perdue held a prayer vigil in Atlanta on November 13, 2007.

40   **our shared academic background:** We are both ethologists—that is, zoologists trained in the study of animal behavior. Dawkins is, moreover, a student of the Dutch father of ethology, Niko Tinbergen, who moved to Oxford in 1947. Tinbergen was a product of the same traditions that shaped my teachers in the Netherlands.

40   **two-level approach:** Biologists distinguish between (1) the reason why a behavior evolved in a species over millions of years, and (2) how individuals produce the behavior in the here and now. We call the first the *ultimate* reason for a behavior's existence, and the second the *proximate* process that produces it (Mayr, 1961; Tinbergen, 1963). The proximate/ultimate distinction is considered one of the toughest in

evolutionary thought, and it is undoubtedly the most violated one. Biologists often focus on the ultimate level at the expense of the proximate one, whereas psychologists do the opposite. Being a biologist with psychological interests, I pursue a proximate perspective (focused on emotion, motivation, and cognition) informed by an evolutionary framework. The idea behind "motivational autonomy" is that motivations behind a behavior are unconstrained by the ultimate reasons for its existence. Even if a behavior evolved for self-serving reasons, these reasons do not need to be part of the actor's motivation any more than that a spider needs to be intent on catching flies while weaving her web.

42   **biologists call such applications a mistake:** In his television documentary, Richard Dawkins struggled with the same issue, speaking of "misfiring selfish genes." He meant that human kindness is applied under a wider range of circumstances than what it originally evolved for, which is another way of saying that it enjoys motivational autonomy. As Matt Ridley (1996, p. 249) put it in *The Origins of Virtue*, "Our minds have been built by selfish genes, but they have been built to be social, trustworthy and cooperative."

42   **jumps on the train tracks:** The evidence for costly altruism in animals is largely anecdotal, but the same is true of our own species. All we can go by are occasional media accounts. Three typical examples:

- Wesley Autrey, a fifty-year-old construction worker, rescued a man who had fallen in front of an approaching New York subway train. Too late to pull him to safety, Mr. Autrey jumped between the tracks, pressed the other man down, and lay on top of him while five cars rolled overhead. Afterward, he downplayed his heroism: "I don't feel like I did something spectacular" (*New York Times*, January 3, 2007).
- In Roseville, California, a black Labrador, named Jet, jumped in front of his friend, six-year-old Kevin Haskell, who was being threatened by a rattlesnake. The dog took the serpent's venom. The boy's family spent four thousand dollars on blood transfusions and veterinary bills to save their pet (KCRA, April 6, 2004). For other costly altruism by a dog, watch the remarkable footage of one who drags an injured companion off a busy Chilean highway: www.youtube.com/watch?v=DgjyhKN_35g.

- Off the coast of New Zealand's North Island, Rob Howes and three other swimmers were herded and protected by dolphins. The dolphins swam tight circles around them, driving them together. When Howes tried to swim free, the two largest dolphins herded him back just as he spotted an approaching nine-foot great white shark. The swimmers remained encircled for forty minutes before the dolphins let them go (New Zealand Press Association, November 22, 2004).

43 **"Scratch an 'altruist'"**: Michael Ghiselin (1974, p. 247).

43 **"pretense of selflessness"**: Robert Wright (1994, p. 344).

43 **Monty Python**: "The Merchant Banker" sketch, *Monty Python's Flying Circus*, 1972.

44 *Chimpanzee Politics*: My book about the political dramas at the Arnhem Zoo focused on power and aggression, drawing parallels with the writings of Niccolò Machiavelli. Nevertheless, in the context of all this jockeying for position, I noticed a great need in the apes to maintain social relationships, make up after fights, and reassure distressed parties, which got me thinking about empathy and cooperation. The death of Luit opened my eyes to the abyss into which these animals fall if conflict management fails.

45 **a backdrop of competition**: Every human society needs to achieve its own equilibrium among three poles: (1) competition over resources, (2) social cohesion and solidarity, and (3) a sustainable environment. Tensions exist between all three poles, but my book focuses exclusively on tension between the first and second ones.

CHAPTER 3: BODIES TALKING TO BODIES

47 **laughter originated from scorn**: Thomas Hobbes (*Leviathan*, 1651, p. 43): "Sudden glory is the passion which maketh those grimaces called LAUGHTER; and is caused either by some sudden act of their own that pleaseth them; or by the apprehension of some deformed thing in another, by comparison whereof they suddenly applaud themselves." A similar view has been expressed by Richard Alexander (1986).

47 **laughing epidemics**: Robert Provine's *Laughter: A Scientific Investigation* (2000) describes *kuru*, a degenerative disease found among cannibals in the highlands of New Guinea. It is marked by excessive laughter

(including laughing at one's own stumbling and falling) even though the disease invariably has a fatal outcome.

48   **when young apes put on their playface:** In a frame-by-frame analysis of orangutans, Marina Davila Ross found involuntary facial mimicry. If one ape showed a playface—even in the absence of any tickling, wrestling, or jumping—its partner would within a second copy the expression (Davila Ross et al., 2007).

48   **"paroxystic respiratory cycle":** From Oliver Walusinski and Bertrand Deputte (2004). Charles Darwin already commented on the yawn as a universal reflex: "Seeing a dog & horse & man yawn, makes me feel how much all animals are built on one structure" (Darwin's Notebook M, 1838). Our own chimpanzee study is being conducted by Matthew Campbell and Devyn Carter. Like other forms of basic empathy, however, catching a yawn is not limited to primates: Human yawns make dogs yawn (Joly-Mascheroni et al., 2008).

49   **copying of small body movements:** Strictly speaking, we don't *copy* yawns, because yawning is an involuntary reflex. All we can say is that yawning by one person *induces* yawns in another. From an interview with Steve Platek: "[T]he more empathetic you are, the more likely it is that you'll identify with a yawner and experience a yawn yourself" (Rebecca Skloot, *New York Times*, December 11, 2005).

50   **baboon troop gathered:** This incident happened in April 2007 at Emmen Zoo, in the Netherlands. Another "mass hysteria" occurred when six newly arrived penguins introduced a strange habit to a colony at the San Francisco Zoo, with all birds swimming in circles for weeks. They started every morning until they staggered totally exhausted out of the pool at dusk. "We've lost complete control," complained the penguin keeper (Associated Press, January 16, 2003).

50   **horses were trapped:** This spectacular horse rescue was set to music on the Internet: www.youtube.com/watch?v=i6vSvOw-4U4.

51   **husky named Isobel:** Struck by blindness, this dog had been removed from her team, but was returned when she stopped eating (*Canadian Press*, November 19, 2007). Isobel's story reminds me of Darwin's (1871, p. 77) account of "an old and completely blind pelican, which was very fat." Darwin speculated that other birds might have been feeding the blind one, but I wonder if it couldn't have been that this bird, like Isobel, accompanied others to feeding grounds relying on hearing and air flow in the tight formations that pelicans are known for. But then the question still remains how a blind bird catches fish.

51 **"Three times when this happened":** Jane Goodall (1990, p. 116).

52 **children with autism:** Eleven-year-old children with autism spectrum disorder yawn as much as other children of the same age, but don't yawn more while watching videotaped yawns, whereas typically developing children show a marked increase (Senju et al., 2007).

53 **neural resonance:** Obviously, mirror neurons could play a role in the copying of mouth movements and facial mimicry (e.g., Ferrari et al., 2003), but this still does not solve the correspondence problem, which requires preexisting knowledge of which body part of another individual corresponds with one's own.

53 **dolphins mimicked people:** Louis Herman (2002) described dolphin imitation, and Bruce Moore (1992) did the same for an African gray parrot. The bird not only mimicked sounds but also body movements. He'd say "Ciao" while waving goodbye with a foot or wing, or say "Look at my tongue" while sticking out his tongue, just as Moore had shown him. This bird thus solved the correspondence problem with a totally different species.

54 **swagger with arms hanging:** White House press release (September 2, 2004) quoting George W. Bush: "Some folks look at me and see a certain swagger, which in Texas is called 'walking.'"

54 **Arthur Miller:** Cited in *Emotional Contagion* by Elaine Hatfield, John Cacioppo, and Richard Rapson (1994, p. 83), which book provides an excellent overview of mimicry and emotional contagion, and is the source of some of my human examples.

54 **Give a zoo ape a broom:** Anne Russon (1996) describes orangutans in a sanctuary imitating human caretakers, such as stringing up hammocks and washing dishes. They also mimic undesirable (i.e., unrewarded) activities such as siphoning gasoline from a drum.

54 **white-coated experimenter:** Apes have traditionally suffered from unfair comparisons with human children, such as when only the apes face a species barrier during testing (e.g., Tomasello, 1999; Povinelli, 2000; Hermann et al., 2007). It is time to move toward ape-to-ape testing, which has greater ecological validity and has produced remarkable breakthroughs in recent years (de Waal, 2001; Boesch, 2007; de Waal et al., 2008).

56 **imitation is a way of reaching a goal:** The classical definition of *imitation* is to learn an act by seeing it done (Thorndike, 1898). This definition covers the term's common meaning—including the my-finger-got-stuck routine described in the text—but narrower definitions

have gained popularity. So-called true imitation entails recognition of another's goal as well as copying of the other's technique to reach this goal (Whiten and Ham, 1992). I prefer the older, broader sense of the term, however, for the simple reason that I believe all forms of imitation to be evolutionarily and neurologically continuous.

56 **Andy Whiten:** A professor of psychology and primatology at St. Andrews University, Whiten developed the two-action paradigm to test ape imitation. He teamed up with our Living Links Center, in Atlanta, to apply this paradigm to group-living chimps. Results strongly support imitation in apes (e.g., Bonnie et al., 2006; Horner and Whiten, 2007; Horner et al., 2006; Whiten et al., 2005), and relate to the ongoing debate about animal "culture" (e.g., de Waal, 2001; Mc-Grew, 2004; Whiten, 2005).

58 **Adult apes are potentially dangerous:** Even a young, relatively small chimpanzee has the muscle strength of several grown men bundled into one. Adult chimpanzees are totally beyond unarmed human control, and have been known to kill people.

59 **a ghost box:** Lydia Hopper demonstrated 225 food deliveries from a box controlled by transparent fish lines, before giving the chimps a chance to manipulate the same box without the lines. They had no clue what to do (Hopper et al., 2007).

59 **ask a pianist to pick out his own performance:** Saying that mental processes run "via our bodies" is shorthand for saying that they run via neural representations and associated proprioceptive sensations of our bodies in our brain. The examples given are perception (Proffitt, 2006) and pianist self-recognition (Repp and Knoblich, 2004).

60 **using a heavy rock as hammer:** Video by Sarah Marshall-Pescini and Andrew Whiten (2008). What I mean by a "shortcut to imitation" is that not all imitation or emulation requires an actual understanding of the other's goals, methods, and rewards. Unconscious motor mimicry bypasses such cognitive appraisals, producing rapid learning based on bodily closeness to the model (cf. Bonding- and Identification-based Observational Learning, or BIOL; de Waal, 2001).

61 **"Once I saw an elephant mother":** Katy Payne (1998, p. 63) in *Silent Thunder.*

62 **When a human experimenter imitates:** Described by Andrew Meltzoff and Keith Moore (1995). Macaques, too, recognize when they are

being imitated (Paukner et al., 2005), and apes even test out the imitator, as human children do (Haun and Call, 2008).

62 **The Dutch may be notoriously stingy:** Dutch restaurant bills include service charges, hence tips are small. They're nevertheless higher for waitresses instructed by scientists to repeat orders (van Baaren et al., 2003).

62 **Like chameleons:** Human copycat tendencies are in fact known as the "chameleon effect" (Chartrand and Bargh, 1999).

64 **"the most complicated opus":** Joe Marshall and Jito Sugardjito (1986, p. 155).

65 **a good siamang marriage:** Thomas Geissmann and Mathias Orgeldinger (2000). The quote comes from an interview in *Spiegel Online* (February 6, 2006). Similar vocal convergence occurs in pairs of male bottlenose dolphins that have formed an alliance: The stronger their bond, the more the males' vocalizations sound alike (Wells, 2003).

65 *Einfühlung:* The terminology came from an earlier German psychologist, Robert Vischer. In the phrasing of Lipps, *Einfühlung* permits us to gain knowledge about the other self (*das andere Ich*) or the foreign self (*das fremde Ich*). See also Schloßberger (2005) and Gallese (2005). The German language is rich in variations on this terminology, from feeling into, feeling with, and suffering with others—each process denoted by its own single word—but also opposites, such as *Schadenfreude* (literally: hurt-joy), that is, getting pleasure out of someone else's pain.

66 **subliminal presentation:** Ulf Dimberg and co-workers (2000). Recent work by Stephanie Preston and Brent Stansfield (2008) shows that the leakage of facial information even includes the conceptual, semantic level.

67 *emotional contagion:* Defined as "the tendency to automatically mimic and synchronize facial expressions, vocalizations, postures, and movements with those of another person and, consequently, to converge emotionally" (Hatfield et al., 1994, p. 5).

67 **one baby crying:** Studies on contagious crying report a stronger response in female than male infants. Some studies have explored a range of other sounds. Human infants respond strongest to real cries produced by other infants, not playbacks of their own cries, cries of older children, chimpanzee screams, or computer-generated wails (Sagi and Hoffman, 1976; Martin and Clark, 1982).

68 **"We haven't yet solved the problem of God":** From Tom Stoppard's (2002) play *The Coast of Utopia*.

70 **One rat's distress:** Joseph Lucke and Daniel Batson (1980) tried to determine if rats are concerned about the companions they give shocks to, and concluded that they are not. This does not deny, of course, that they can be emotionally affected by another's distress.

71 **The last mouse showed more signs:** Jeffrey Mogil on National Public Radio (July 5, 2006). The study on commiserating mice was published by Dale Langford et al. (2006).

73 **Oscar the Cat:** David Dosa (2007), a geriatrician, published "A Day in the Life of Oscar the Cat," saying: "His mere presence at the bedside is viewed by physicians and nursing home staff as an almost absolute indicator of impending death, allowing staff members to adequately notify families. Oscar has provided companionship to those who otherwise would have died alone" (p. 329).

75 **avoid unpleasant sights and sounds:** Inasmuch as self-protective altruism seeks to reduce negative arousal caused by the state another finds itself in, it is based on empathy. I am using the altruism label here in the biological sense: behavior that benefits another at a cost to the self regardless of whether the effect on the other is intended (chapter 2).

76 **"a much more skilled interpreter":** Quoted from Robert Miller (1967, p. 131).

76 **I avoid causing pain:** The ethics of animal research is subject to never-ending, often acrimonious debate. Since my own research doesn't aim at solving pressing medical problems, I feel there is little justification for invasive procedures. My personal two rules of thumb are that (1) I work only with group-living (as opposed to singly housed) primates, and (2) I use relatively stress-free procedures, defined as procedures that I wouldn't mind applying to human volunteers.

77 **grooming slows down the heart:** The monkey project was led and published by Filippo Aureli and co-workers (1999). A heart-rate study on geese by Claudia Wascher and co-workers (2008) found that birds implanted with transmitters would get emotionally aroused by the mere sight of their mate being in trouble with others, thus suggesting emotional contagion in birds.

78 **empathy literature is completely human-centered:** A notable exception was psychologist William McDougall (1908, p. 93), who did recognize empathy in gregarious animals, offering us the following

insightful characterization of empathy: "The cement that binds all animal societies together, renders the actions of all members of a group harmonious, and allows them to reap some of the prime advantages of social life."

78 **automatic reactivation of neural circuits:** Empathy rests on a property of the nervous system that (1) activates its own neural substrates for emotion and action upon perceiving emotions and actions in others, and (2) uses these activated states within the self to access and understand the other. This idea goes back to Lipps's (1903) writing on *innere Nachahmung* (i.e., inner mimicry). Stephanie and I reformulated this as the perception-action mechanism of empathy (Preston and de Waal, 2002). Even while merely imagining another's situation, humans automatically activate these neural substrates. Thus, when subjects are asked to put themselves into another's shoes, their brain activation is similar to when they recall similar situations that involved themselves (Preston et al., 2007).

79 **Pink Floyd:** In "Echoes," on the album *Meddle* (1971). Band member Roger Waters noted in an interview: "[It] has a lyric about strangers passing on the street that's become a recurrent theme for me, the idea of recognizing oneself in others and feeling empathy and a connection to the human race" (*USA Today*, August 6, 1999).

79 **The discovery of mirror neurons:** Vilayanur Ramachandran: "I predict that mirror neurons will do for psychology what DNA did for biology: they will provide a unifying framework and help explain a host of mental abilities that have hitherto remained mysterious and inaccessible to experiments" (Edge.org, June 1, 2000). How exactly mirror neurons translate into imitation and empathy remains unclear, however, but see Vittorio Gallese and co-workers (2004), and Marco Iacoboni (2005). Mirror neurons have also been found in birds, so that the perception-action mechanism probably goes all the way back to the shared reptilian ancestor of mammals and birds (Prather et al., 2008).

79 **empathize with everybody:** Commentaries on Preston and de Waal (2002).

80 **Identification is such a basic precondition:** In the monkey experiments mentioned before, too, familiarity enhanced empathic responses (Miller et al., 1959; Masserman et al., 1964).

80 **when groups compete:** For in-group biases in empathy, see Stefan Stürmer and co-workers (2005).

80   **"dechimpized":** Jane Goodall (1986, p. 532).

81   **entire body expresses emotions:** Rhesus monkeys avoid pictures of conspecifics in a fearful pose, which arouse a stronger response than negatively conditioned stimuli (Miller et al., 1959).

81   **the body posture won out:** With emotionally congruent pictures (i.e., face and body express the same emotion), the reaction time was on average 774 milliseconds, whereas with emotionally incongruent pictures (i.e., face and body express opposite emotions) it was 840 milliseconds, both still under one second (Meeren et al., 2005).

82   **Emotional contagion thus relies:** Beatrice de Gelder (2006) contrasts the Body First Theory (also: the James-Lange theory) with the Emotion First Theory. The latter rests on two closely integrated levels: a fast, reflexlike process, not unlike the perception-action mechanism, and a slower, more cognitive appraisal of stimuli in context.

82   **the face remains the emotion highway:** The face is the seat of individual identity. Whom we are dealing with determines identification, which in turn affects our reactions.

83   **empathy needs a face:** This felicitous phrase as well as the example of Parkinson's patients come from Jonathan Cole (2001).

83   **"I live in the facial expression":** Maurice Merleau-Ponty (1964, p. 146).

83   **"I have returned to the planet":** The anonymous woman with the face transplant: "Je suis revenue sur la planète des humains. Ceux qui ont un visage, un sourire, des expressions faciales qui leur permettent de communiquer" ("La Femme aux Deux Visages," *Le Monde,* June 7, 2007).

CHAPTER 4: SOMEONE ELSE'S SHOES

84   **"Sympathy . . . cannot in any sense":** Adam Smith (1759, p. 317).

84   **"Empathy may be uniquely well suited":** Martin Hoffman (1981, p. 133).

84   **Nadia Kohts:** Her full name was Nadezhda Nikolaevna Ladygina-Kohts. She lived from 1889 to 1963 and was the wife of Aleksandr Fiodorovich Kohts, founding director of Moscow's State Darwin Museum.

85   **among the stuffed animals in the basement:** In 2007, Moscow's Darwin Museum celebrated its hundredth anniversary, displaying historical photographs that the staff had shown me of Kohts doing her pioneering research. Apart from her work with Yoni and other primates, I saw pictures of her accepting an object handed to her by a

large cockatoo, and her holding out a tray with a choice of three cups toward a macaw. Her tests had a distinctly modern look, and Kohts often had a smile on her face, evidently liking her work. She tested ape tool use at the same time as Wolfgang Köhler, and may be the discoverer of the matching-to-sample technique still universally applied in visual cognition research. The only book (out of seven) by Kohts translated into English is *Infant Chimpanzee and Human Child* (2002), originally published in Russian in 1935.

86   **"If I pretend to be crying"**: Ladygina-Kohts (1935, p. 121).

88   **"The definition of sympathy"**: Lauren Wispé (1991, p. 68).

89   **Abraham Lincoln:** The story goes that Lincoln halted his carriage to attend to a squealing pig mired in the mud, and dragged it out while soiling his good pants. There even exists a children's book, *Abe Lincoln and the Muddy Pig* (Krensky, 2002).

89   **Good Samaritan:** This experiment on constraints on human sympathy has become a classic. It was conducted by John Darley and Daniel Batson (1973).

90   **thousands of consolations:** Consolation behavior is so common in apes (de Waal and van Roosmalen, 1979) that at least a dozen studies now offer quantitative details. Recently, Orlaith Fraser and co-workers (2008) confirmed that consolation has a stress-reducing effect on its recipients. The large-scale analysis referred to in the text is being conducted by M. Teresa Romero on our computer records of more than two hundred thousand spontaneous social events among chimpanzees.

91   **"Impressive indeed is the thoughtfulness"**: The quote is from Robert Yerkes (1925, p. 131). Yerkes was so struck by the concern shown by Prince Chim for Panzee, his terminally ill companion, that he admitted, "If I were to tell of his altruistic and obviously sympathetic behavior towards Panzee I should be suspected of idealizing an ape" (p. 246).

91   **a little duckling:** Peter Bos (personal communication).

92   **man's best friend:** The Belgian study was conducted by Anemieke Cools and co-workers (2008).

93   **The ancestor of the dog:** For wolves we do not as yet have the same evidence on consolation (i.e., reassurance of a distressed party by a bystander), but there are observations of wolf reconciliation (i.e., a reunion between former opponents) by Giada Cordoni and Elisabetta Palagi (2008).

93   **"We are about to die"**: Anthony Swofford (2003, p. 303).

94   **famous images:** A 1950 photograph by Al Chang, which inspired my drawing.

94   **nightmare of losing a child:** Interview with Paul Rosenblatt by Kate Murphy (*New York Times*, September 19, 2006).

94   **"no-hug policy":** "School Enforces Strict No-Touching Rule" (Associated Press, June 18, 2007).

94   **young rhesus monkeys:** For a decade, I studied two large troops of rhesus macaques at the Vilas Park Zoo in Madison, Wisconsin. Rhesus are seasonal breeders. Every spring, about twenty-five infants were born at about the same time. This created a mass of same-age peers, which were very much in tune with one another in playfulness, sleepiness, and distress (de Waal, 1989).

96   **It develops out of a primitive urge:** Nature is full of inborn tendencies that help members of a species acquire critical skills. For example, capuchin monkeys are born with an insuppressible tendency to bang small objects that they can't open, which they'll do with gusto for hours. Cats have been endowed with an insuppressible tendency to lock their eyes onto any moving object small enough to pounce on. Combined with experience and learning, such tendencies are gradually incorporated into skills such as nut cracking with stones, which capuchins do in the wild (Ottoni and Mannu, 2001), or stalking and hunting, which all cats do. Preconcern is another inborn tendency that promotes further learning.

96   **Most mammals show some:** Empathy is like a multilayered Russian doll with the ancient perception-action mechanism and emotional contagion at its core, around which ever greater complexities have been constructed (de Waal, 2003; chapter 7).

97   **"If Rock was not present":** Emil Menzel (1974, pp. 134–135).

97   *theory of mind...:* Emil Menzel's work (e.g., Menzel, 1974; Menzel and Johnson, 1976) combined with Nicholas Humphrey's (1978) notion of animals as "natural psychologists" (i.e., modeling the other's mind) preceded or coincided with David Premack and Guy Woodruff's (1978) influential "theory of mind" concept, published a few years after Menzel started working with Premack at the University of Pennsylvania. Theory of mind refers to the ability to recognize the mental states of others.

98   **champion mind readers:** Since Maxi's belief is incorrect, this is known as the "false belief" task. This task relies so heavily on language, though, that language skills affect its outcome. If the role of language

is reduced, children of a younger age show evidence of understanding beliefs, suggesting simpler processes than hitherto assumed (Perner and Ruffman, 2005).

99   **Ravens have large brains:** From an interview with Thomas Bugnyar in *The Economist* (May 13, 2004). See Bugnyar and Bernd Heinrich (2005). Further evidence for perspective-taking in birds was provided by Joanna Dally and co-workers (2006).

100   **mental states of others:** The first experimental dent in claims that only humans possess theory-of-mind came from a study with our chimps at the Yerkes Primate Center by Brian Hare and co-workers (2001). They showed that low-ranking apes take the knowledge of a dominant competitor into account before approaching food. Further successful ape studies have been reviewed by Michael Tomasello and Josep Call (2006), but note also evidence for perspective taking in birds (above), dogs (Virányi et al., 2005), and monkeys, (Kuroshima et al., 2003; Flombaum and Santos, 2005).

100   **"changing places in fancy":** Adam Smith's (1759, p. 10) classical description referred to sympathy. Cold perspective-taking, on the other hand, may be closer to what is commonly known as theory-of-mind, even though the word "theory" falsely implies abstract thinking and extrapolation from self to other by means of reasoning, for which there is no evidence (de Gelder, 1987; Hobson, 1991). More likely, perspective-taking develops out of the sort of unconscious bodily connections discussed in chapter 3.

100   **When a juvenile orangutan:** Reported in *The Sydney Morning Herald* (February 14, 2008).

100   **Swedish zoo:** The chimp-and-rope incident occurred at Furuvik Park, in Gavle, and was described to me by the primate curator, Ing-Marie Persson.

101   **I snared Emil for an interview:** Emil was born in 1929. The interview took place in 2000. A few years later, one of his former students wrote me: "I am presently a professor of developmental psychology. Once, on my way to the greenhouse where we kept our marmoset colony, I had to walk through a hallway where Emil's chimps had been let out, and were roaming around. I was somewhat fearful of walking past them, and the young one, Kenton, walked up to me and gently took my hand, leading me through the hallway past the other chimps. I observed chimps' capacity for empathy firsthand!" (Alison Nash, personal communication).

102 **Since all animals rely:** The lecture took place at Wesleyan College, and the overbearing chairman was Richard Herrnstein (1930–94), one of the foremost Skinnerians at the time. Herrnstein felt that pigeons could easily take the place of chimpanzees, similar to B. F. Skinner's opinion: "Pigeon, rat, monkey, which is which? It doesn't matter" (Bailey, 1986).

102 **a spectacular escape:** The chimpanzee escape was published as "Spontaneous Invention of Ladders in a Group of Young Chimpanzees" by Menzel (1972). In *Chimpanzee Politics,* I describe a very similar cooperative escape (de Waal, 1982).

104 **"Mother Number One":** "Officer Breast-Feeds Quake Orphans" (CNN International, May 22, 2008).

104 **hominid was recently found in the Caucasus:** A 1.8-million-year-old fossil discovered by David Lordkipanidze and colleagues (2007).

104 **Madame Bee:** Jane Goodall (1986, p. 357).

105 **handful of stories:** The best-known anecdote, captured on video, is the rescue of a human child at the Brookfield Zoo in Chicago. On August 16, 1996, Binti Jua, an eight-year-old gorilla, saved a three-year-old boy who had fallen eighteen feet into the primate exhibit. The gorilla sat down on a log in a stream, cradling the boy in her lap, giving him a gentle back-pat before she continued on her way. This act of sympathy touched many hearts, making Binti a celebrity overnight (*Time* elected her one of the "Best People" of 1996). The number of similar anecdotes keeps growing. I have been trying not to repeat stories used before, such as in *Good Natured* (1996) and *Bonobo* (1997). A systematic overview has been put together by Sanjida O'Connell (1995).

105 **Chimp Haven:** The relation between chimpanzees Sheila and Sara was described to me by Amy Fultz, who works at Chimp Haven, located near Shreveport, Louisiana. Amy also described one chimp going out of her way to bring food to another who was incapacitated by kidney disease. For more on Chimp Haven (with which I am involved) and how to support it, see www.chimphaven.org.

106 **apes on islands:** The suggestion that animals never take serious risks on behalf of one another was made by Jeremy Kagan (2000). Examples of apes jumping into water to save another come from Jane Goodall (1990, p. 213) and Roger Fouts (1997, p. 180), including a rescue by Washoe of another female that she had known for only a few hours. The mother-son drowning occurred at the Dublin Zoo (*Belfast News Letter,* October 31, 2000).

100 **hydrophobia cannot be overcome:** Helping behavior may have evolved in the context of kinship and reciprocity, but there is little evidence that chimpanzees actually count on return favors (chapter 6). Even for humans, who *are* capable of such anticipation, it is questionable that anyone would run into a burning building or jump into water with return favors in mind. The impulse is likely emotional. Again, the actor's reasons for a behavior do not need to overlap with the reasons for its evolution, which may indeed be self-interested (chapter 2).

107 **leopard attack:** Christophe Boesch (personal communication) has documented regular predation on chimpanzees in Ivory Coast. Chimps help one another against leopard attacks, thus taking grave risks on behalf of one another.

108 **Children read "hearts":** Children pass the traditional theory-of-mind tasks, which focus on beliefs, around the age of four. But they appreciate the feelings, needs, and desires of others much earlier, usually at the age of two or three (Wellman et al., 2000). The trouble older children have with the Little Red Riding Hood story seems due to emotional identification, which interferes with the attribution of beliefs (Bradmetz and Schneider, 1999).

110 **Social scratching:** Primate customs and traditions, also known as "cultural primatology," are the subject of *The Ape and the Sushi Master* (de Waal, 2001). For details on the social scratch of the Mahale chimpanzees, see Michio Nakamura and co-workers (2000).

112 **understand when one among them is hungry:** There is little evidence that monkeys appreciate the knowledge or beliefs of others, but this doesn't keep them from appreciating another's attention, intentions, or needs. Conducted by Yuko Hattori, our food-sharing experiment tested responses to partners who had just eaten, or not, and included a control condition in which the partner had been behind an opaque panel, so that subjects could not know about its previous food consumption.

113 **monkeys favor sharing:** The prosocial choices in capuchins disappeared if the partner was either a stranger or out of sight (de Waal et al., 2008; chapter 6). Similar prosocial preferences have been demonstrated in monkeys by Judith Burkart and co-workers (2007) and Venkat Lakshminarayanan and Laurie Santos (2008).

113 **"Chimpanzees Are Indifferent":** This is the actual title of a scientific article by Joan Silk and co-workers (2005). A similar outcome was

reported by Keith Jensen and co-workers (2006). It is almost impossible to interpret negative findings, however (de Waal, 2009). A common problem is that animals may fail to fully understand the task. If they develop a blind routine, for example, or are too far apart to notice what happens to their partner, their choices may appear socially indifferent yet are in fact better described as statistically random.

115 **rewards made no difference:** Felix Warneken and co-workers (2007) included conditions with and without rewards. Since these conditions had no effect, the chimps' helping behavior did not seem to be driven by expected payoffs.

115 **"His euphoria produced":** Quoted from Dolf Zillmann and Joanne Cantor (1977, p. 161). See also Lanzetta and Englis (1989).

116 **how altruistic is altruism:** This has been explored empirically in Daniel Batson's (1991, 1997) admirable work on the self- versus other-orientation behind human altruism. The debate about this issue is never-ending, though, because of the impossibility to extract the self from its relations with others, especially with regard to empathy (e.g. Hornstein, 1991; Krebs, 1991, Cialdini et al., 1997).

CHAPTER 5: THE ELEPHANT IN THE ROOM

118 **"Seeing himself in the mirror":** Ladygina-Kohts (1935, p. 160).

119 **Pliny the Elder:** From his *Natural History* (vol. 3, Loeb Classical Library, 1940).

121 **predicted decades ago:** In 1970, Gordon Gallup, Jr., published his first study of mirror self-recognition (MSR), followed by speculations a decade later on how MSR correlates with other so-called "markers of mind," including attribution and empathy. Gallup (1983) explicitly speculated that cetaceans and elephants show enough insightful social behavior that they probably also possess MSR.

122 **arrive in drag:** As an undergraduate student, I worked with two young male chimpanzees. A male fellow student and myself wanted to know why these apes were sexually aroused by every woman in sight (e.g., secretaries, students), and especially how they told the human genders apart. So, we dressed up in drag and changed the pitch of our voices. But the chimps were not confused, least of all sexually.

123 *co-emergence hypothesis:* This hypothesis has its origin in the separate perspectives of Gordon Gallup on phylogeny and Doris Bischof-Köhler on human ontogeny, both of which link mirror responses to

social cognition. My own contribution is to combine these two perspectives into a single hypothesis.

123 **When the same children:** Co-emergence in development of personal pronoun use, pretend play, and mirror self-recognition (MSR) was demonstrated by Michael Lewis and Douglas Ramsay (2004). Doris Bischof-Köhler has conducted the most detailed studies of the co-emergence of MSR and empathy in children, suggesting an absolute link; that is, "empathizers" pass the rouge test in front of the mirror, whereas "non-empathizers" fail this test. This connection persists after correction for age (Bischof-Köhler, 1988, 1991) and has also been reported by Johnson (1992) and Zahn-Waxler et al. (1992).

124 **neuroscience will one day resolve:** Neuroimaging studies on the role of the self in empathy are under way, following the ideas of Jean Decety (Decety and Chaminade, 2003). Advanced empathy is likely based on the perception-action mechanism combined with an increasing self-other distinction (Preston and de Waal, 2002; de Waal, 2008). In humans, the right inferior parietal cortex, at the temporoparietal junction (TPJ), helps distinguish self- from other-produced actions (Decety and Grèzes, 2006).

124 **"Self-absorption kills empathy":** Daniel Goleman in *Social Intelligence* (2006, p. 54).

125 **ontogeny and phylogeny:** Even though modern biology rejects Ernst Haeckel's recapitulation theory, it remains true that if an anatomical feature evolved before another, it generally also develops earlier in the embryo, and that shared ancestry among species is often reflected in the early stages of embryonic development. The co-emergence hypothesis of mirror self-recognition and social cognition draws a parallel between ontogeny and phylogeny without implying an obligatory connection between the two. See also Gerhard Medicus (1992).

126 **Even a goldfish will jump:** In an interview, Paul Manger claimed: "You put an animal in a box, even a lab rat or gerbil, and the first thing it wants to do is climb out of it. If you don't put a lid on top of the bowl of a goldfish it will eventually jump out to enlarge the environment it is living in. But a dolphin will never do that. In the marine parks the dividers to keep the dolphins apart are only a foot or two above the water" (Reuters, August 18, 2006). Manger did not speculate that it may actually be smart for an animal to stay in a known environment rather than jump to an unknown one.

126 **preen themselves before a mirror:** Mirror self-recognition is not a trained pet trick but rather a spontaneous capacity that some animals possess and others don't. Training the criterion behavior (cf. Epstein et al., 1981) defeats the purpose of the rouge test and can only produce the sort of "trivial passing" that machines are also capable of. Moreover, when another research team tried to replicate the pigeon study, they failed miserably, resulting in a paper with the word "Pinocchio" in its title (Thompson and Contie, 1994).

127 **Dolphins possess large brains:** The human brain weighs approximately 1.3 kilograms, the bottlenose dolphin's 1.8, the chimpanzee's 0.4, and the Asian elephant's 5. If brain size is corrected for body size, the human brain is larger than that of any other animal, and cetacean brains are larger than those of nonhuman primates (Marino, 1998). Some analyses stress different parts of the brain, but in this regard human uniqueness is less striking. Contrary to general belief, the human frontal cortex is no larger than that of the great apes relative to the rest of the brain (Semendeferi et al., 2002).

127 **rightly upset dolphin experts:** Manger's (2006) article provoked a collective rebuttal by many of the world's dolphin experts in an article titled "Cetaceans Have Complex Brains for Complex Cognition" by Lori Marino et al. (2007).

128 **a stick of dynamite:** Drawing based on J. B. Siebenaler and David Caldwell (1956).

128 **Reports of leviathan care:** Further examples have been provided by Melba Caldwell and David Caldwell (1966) and Richard Connor and Kenneth Norris (1982).

129 **nudged to shore by a seal:** "Seal Saves Drowning Dog" (BBC News, June 19, 2002).

129 **a female humpback whale . . . :** Incident reported by Peter Fimrite in *The San Francisco Chronicle* (December 14, 2005). Even though the report adds a condescending disclaimer ("Whale experts say it's nice to think that the whale was thanking its rescuers, but nobody really knows what was on its mind") it should be noted that for species with complex reciprocity, gratitude is an expected emotion (Trivers, 1971; Bonnie and de Waal, 2004).

130 **website for a conference:** "What Makes Us Human," held in April 2008 in Los Angeles.

130 **"I always smile when I hear Garrison Keillor":** From Michael Gazzaniga, "Are Human Brains Unique?" (*Edge,* April 10, 2007). The author

answers his own question as follows: "Something like a phase shift has occurred in becoming human. There simply is no one thing that will ever account for our spectacular abilities." The vagueness of this answer amounts to an admission that the human brain is in fact *not* that unique.

132 **six blind men from Indostan:** From the poem "Blind Men and the Elephant" by John Godfrey Saxe, published in 1873.

132 **"Eleanor was found with a swollen trunk":** The incident occurred on October 10, 2003, and was recorded and photographed by Iain Douglas-Hamilton and co-workers (2006).

133 **a poacher's bullet:** Example from Cynthia Moss's (1988, p. 73) *Elephant Memories*. The bull spraying another with water is described in *African Elephants, A Celebration of Majesty* by Daryl and Sharna Balfour (1998), and the mud-hole scene was shown on *National Geographic*'s show *Reflections on Elephants* (1994). For a review of empathy-related behavior in African elephants, see Lucy Bates and co-workers (2008).

135 **bigger than an earlier study:** A drawing illustrates the setup of Daniel Povinelli's (1989) elephant experiment.

138 **absent in all other primates:** Esther Nimchinsky and co-workers (1999) compared the brains of twenty-eight primate species, finding VEN cells only in the four great apes and humans. Only one bonobo specimen was available; its brain showed the most humanlike density and distribution of VEN cells, which is intriguing in relation to this species' hypothetical status as the most empathic ape (de Waal, 1997).

138 **special kind of dementia:** Human patients with frontotemporal dementia were studied by William Seeley et al. (2006), who found that three-quarters of all VEN cells in the anterior cingulate cortex were lost in these patients.

138 **not limited to humans and apes:** Thus far, the connection seems tight: All mammals with MSR have VEN cells, and vice versa. But exceptions may yet be found, and the precise function of these cells remains a mystery. For VEN cells in nonprimates, see Atiya Hakeem and co-workers (2009).

140 **"She appears genuinely concerned":** Dorothy Cheney and Robert Seyfarth (2008, p. 156) hint at emotional arousal in baboons without empathic perspective-taking.

142 **attacked a camel herder:** Joyce Poole in *Coming of Age with Elephants* (1996, p. 163).

143 **bereaved baboons:** Study by Anne Engh and co-workers (2005), discussed in "Baboons in Mourning Seek Comfort Among Friends" (ScienceDaily.com, January 31, 2006).

143 **"nightmare of anxiety":** Eugène Marais, a South African naturalist, published *My Friends the Baboons* in 1939.

143 **The monkey had given only a tiny nip:** John Allman (personal communication). Even if capuchins have a capacity for consolation, research does not always demonstrate it because we compare average postconflict and baseline data. But Peter Verbeek did find evidence that capuchin victims of aggression seek contact with others and are treated in an exceptionally friendly manner (Verbeek and de Waal, 1997).

144 **hamadryas baboons:** Drawing based on a photograph in Hans Kummer's *Social Organization of Hamadryas Baboons* (1968, p. 60).

144 **incidents in wild bonnet monkeys:** Anindya Sinha (personal communication).

145 **baboons sometimes reassure distressed infants:** Example from Barbara Smuts (1985, p. 112), who adds that the male's grunting was particularly striking as the infant showed no agony: "Achilles behaved as if he thought that slipping in the sand was an experience that deserved the same sort of reassurance he would normally give when the infant of a female friend screamed or geckered in distress."

145 **"One day, as she leapt":** From Robert Sapolsky's *A Primate's Memoir* (2001, p. 240).

146 **Ahla, a baboon:** Baboons used to be employed as goatherds in South Africa, often riding one of the larger goats, as described by Walter von Hoesch (1961). Cheney and Seyfarth (2008, p. 34) cite an ex-owner, according to whom baboons require no special training to become experts at goat mother-kid relationships.

146 **bridging behavior:** Filippo Aureli and Colleen Schaffner (personal communication) on wild spider monkeys. The earliest descriptions of bridging behavior and its implications came from Ray Carpenter (1934). Further see Daniel Povinelli and John Cant's (1995) attempt to link arboreal locomotion to the self-concept.

147 **monkeys . . . don't pass:** Mirror research on primates has been reviewed by James Anderson and Gordon Gallup (1999).

147 **The self is part of every action:** The obligatory self-awareness of all animals is discussed by Emanuela Cenami Spada and co-workers (1995) and Mark Bekoff and Paul Sherman (2003). Animals lacking mirror

self-recognition nevertheless often understand self-agency (Jorgensen et al., 1995; Toda and Watanabe, 2008).

148 **mirror understanding:** Instead of splitting species into those that recognize themselves in a mirror and those that don't, there exist intermediate levels of understanding. Some animals, such as budgies or fighter fish, never stop courting or fighting their mirror image, whereas most dogs and cats at least gradually lose interest in mirrors. Capuchin monkeys go further in that they seem to perceive their reflection right away as different from a real monkey (de Waal et al., 2005). A gradualist perspective on mirror understanding also typifies studies of young children (Rochat, 2003).

150 **"it takes a thief to know a thief":** As Brandon Keim amusingly put it: "First comes self-knowledge, then crime: it's like a Garden of Eden myth for birds!" (*Wired,* August 19, 2008). Obviously, corvids may use perspective-taking not just for deception, but also for helping and/or consolation, for which there is indeed some evidence (Seed et al., 2007). Further see Nathan Emery and Nicky Clayton's studies of these fascinating birds (2001, 2004). The obvious question of whether magpies or other corvids have VEN cells (like mammals with MSR) may not be relevant: The architecture of the bird brain is so different that similar capacities as in mammals likely came about through convergent evolution, and hence do not necessarily share the same neural substrate.

152 **the hunting dog:** Examples from Susan Stanich (personal communication).

153 **"a bipedal animal such as man":** Quoted from an unpublished 1974 manuscript by Emil Menzel. The author offers extraordinarily detailed descriptions of how chimpanzees deduce the nature and location of hidden objects from the behavior of knowledgeable others, concluding that chimps "have a very effective system of directional communication, to which nothing would be added by manual signalling."

154 **spit into the grass:** This example argues against explanations based on human modeling or training, because no one ever taught Liza to spit for grapes.

154 **"The threatened female":** From *Chimpanzee Politics* (de Waal, 1982, p. 27).

155 **"Noises are heard":** Joaquim Veà and Jordi Sabater-Pi (1998, p. 289).

155 **One difference with human pointing:** Michael Tomasello considers declarative pointing a typically human part of language development:

"Apes are not motivated to simply share information and attitudes with others, nor do they comprehend when others attempt to communicate with these motives" (Tomasello et al., 2007, p. 718). The examples that I offer in the text—pointing at hidden scientists, showing off stinky maggots—contradict this view, suggesting that the volunteering of information is not entirely absent in apes. It remains true, however, that they are less inclined than humans to engage in such behavior.

CHAPTER 6: FAIR IS FAIR

158 **"Every man is presumed to seek":** From Thomas Hobbes's *De Cive* (1651, p. 36).

159 **"He looked at his beautiful hands":** Irène Némirovsky (2006, p. 35).

161 **attached to a dancing pole:** Nigel Scullion, a senior Australian politician, was arrested in a Russian strip club (*Skynews*, December 12, 2007).

161 **Present-day egalitarians:** Egalitarianism is hard work, as Christopher Boehm's (1993) studies make clear. The basic human tendency is social stratification, but in many small-scale societies people actively employ "leveling mechanisms" to keep ambitious males from gaining control. This kind of political organization may have typified much of human prehistory.

161 **Sigmund Freud:** In *Totem and Taboo*, Freud (1913) described "Darwin's primal horde," in which a jealous, violent father kept all women for himself, driving out his sons as soon as they were grown.

161 **the same areas in men's brains:** Study by Brian Knutson and co-workers (2008). The quote is from Kevin McCabe in "Men's Brains Link Sex and Money" (CNN International, April 12, 2008).

162 **"What we think about ourselves":** From Robert Frank's *Passions within Reason* (1988, p. xi). Frank was one of the first to claim that traditional self-interest models fail to account for many aspects of human economic activity.

163 **bringing home stag:** Brian Skyrms (2004).

163 **perished on K2:** "The Descent of Men," by Maurice Isserman (*New York Times*, August 10, 2008).

164 **eye-poking:** The eye-poking game is described by Susan Perry and co-authors (2003). See also Perry (2008).

167 **In one experiment:** Experiment described by Toh-Kyeong Ahn and co-authors (2003). For literature that questions "rational choice" models in economy, see Herbert Gintis's co-edited *Moral Sentiments*

and Material Interests (2005), Paul Zak's *Moral Markets* (2008), Michael Shermer's *The Mind of the Market* (2008), and Pauline Rosenau (2006).

168 **Williams syndrome:** Ursula Bellugi and co-workers (2000). The child's quote is from David Dobbs, "The Gregarious Brain," *New York Times Magazine*, July 8, 2007.

170 **Hermit crabs:** Ivan Chase (1988).

170 **"Nobody ever saw a dog":** From Adam Smith's (1776) *The Wealth of Nations.*

172 **"Let us take a group of volunteers":** Petr Kropotkin (1906, p. 190), *The Conquest of Bread.*

172 **vampire bats:** Gerald Wilkinson (1988).

173 **partake in the hunt:** Contingency between a chimpanzee's contribution to the hunt and access to meat has been suggested by Christophe and Hedwige Boesch (2000).

173 **hierarchy takes a backseat:** It is remarkable how little effect the dominance hierarchy has when apes enter food-sharing mode. This has been remarked upon by fieldworkers, and is well documented in captivity (de Waal, 1989). Primatologists speak of "respect of possession"—that is, once an adult of any rank has become the owner of an item, others give up their claims (e.g., Kummer, 1991).

174 **Socko and May:** Our food-for-grooming study involved a huge computerized database of spontaneous services. Sequential analyses showed chimpanzees capable of memory-based reciprocal exchange (de Waal, 1997).

175 **"marketplace of services":** In *Chimpanzee Politics* (1982), I proposed that apes trade a wide range of services, from grooming and support to food and sex. Ronald Noë and Peter Hammerstein (1994) formulated their biological market theory, which postulates that the value of commodities and partners varies with their availability. This theory applies whenever trading partners can choose whom to deal with. The baboon baby market is one of a growing number of illustrations (Henzi and Barrett, 2002).

176 **capuchins share the meat:** For capuchin monkey hunting and meat sharing see Susan Perry and Lisa Rose (1994) and Rose (1997).

177 **taught to bottle-feed:** The adoption of Roosje and the eternal gratitude of her adoptive mother, Kuif, is detailed in *Our Inner Ape* (de Waal, 2005, p. 202).

178 **poacher's snare:** Stephen Amati and co-workers (2008) describe the damage caused by snares, sometimes leading to the loss of limbs. One

male chimpanzee removed a nylon snare around a female's hand by carefully inspecting it before biting through it.

179 **raid surrounding papaya plantations:** Forward-looking political or sexual exchanges among apes have been documented by de Waal (1982), Nishida et al. (1992), and Hockings et al. (2007). Kimberly Hockings was quoted by *ScienceDaily* (September 14, 2007).

179 **Chester Zoo:** Nicola Koyama and co-workers (2006).

179 **human cooperation a "huge anomaly":** Ernst Fehr and Urs Fischbacher (2003) open their article on altruism with "Human society represents a huge anomaly in the animal world." The reason they give is that humans cooperate with nonrelatives, whereas animals restrict cooperation to close kin, as also in the following characterization by Robert Boyd (2006, p. 1555): "The behavior of other primates is easy to understand. Natural selection only favors individually costly, prosocial behavior when the beneficiaries of the behavior are disproportionately likely to share the genes that are associated with the behavior."

180 **genotyping project:** Kevin Langergraber and co-workers (2007, p. 7788) contradicted the view that primates only help kin, concluding about wild chimpanzees: "males in the majority of highly affiliative and cooperative dyads are unrelated or distantly related." At the Arnhem Zoo, too, unrelated male chimpanzees formed close partnerships, taking serious risks on behalf of one another (de Waal, 1982). Humanity's other closest relative, the bonobo, is marked by high levels of female solidarity resulting in collective dominance over males. Since females are the migratory sex, they lack close genetic ties within any bonobo community, which is why they are said to form a "secondary sisterhood" (de Waal, 1997). This is another example of large-scale cooperation among nonrelatives.

181 **Chimps settle scores:** Primate retribution and revenge was first statistically demonstrated by de Waal and Luttrell (1988) and is further described in *Good Natured* (de Waal, 1996).

182 **"If humans show":** Robert Trivers (2004, p. 964).

182 **the true cradle of cooperation:** The burgeoning literature on strong reciprocity (SR) postulates human prosocial tendencies and punishment of noncooperators. Such behavior is indeed well documented (Herbert Gintis and co-editors, 2005), but there is debate whether SR evolved in order to deal with anonymous outsiders, a category not typically considered in evolutionary models. SR is, in fact, easier un-

derstood as having started within the community, after which it was generalized to outsiders (Burnham and Johnson, 2005).

182 **Bob Dylan's observation:** From the 1983 song "License to Kill."

183 **the king's wives:** While much of the country lives on food aid, nine of the king's thirteen wives went shopping overseas (BBC News, August 21, 2008).

183 **"So the last will be first":** The parable of the workers in the vineyard (Matthew 20:1–16) is not so much about monetary rewards as it is about entering the Kingdom of Heaven. The parable works, however, because of our sensitivity to its fairness aspect.

184 **The chief emotions are egocentric:** Selfish considerations behind the desire for fairness were investigated via the Dictator Game by Jason Dana and co-workers (2004).

185 **French Enlightenment:** Elisabetta Visalberghi and James Anderson (2008, p. 283): "Valuing fairness to others is a rather recent human moral principle, at least in Western cultures, grounded in the theoretical stance—expressed by French Enlightenment philosophers—that people are equal."

186 **Brain scans of players:** Invented by Werner Güth, the ultimatum game was used in an influential study by Daniel Kahneman and co-workers (1986). Alan Sanfey and co-workers (2003) scanned the brains of players facing low offers, and found activation of areas associated with negative emotions.

186 **Lamalera whale hunters:** The hunters collect a small number of whales per year for their own subsistence. They row towards the whale, then the harpoonist jumps on its back to thrust its weapon into it. They follow the whale for hours, often losing it, or killing it through blood loss and exhaustion (Alvard, 2004).

187 **"inequity aversion":** Ernst Fehr and Klaus Schmidt (1999).

188 **Richard Grasso:** James Surowiecki's "The Coup de Grasso," *New Yorker*, October 5, 2003.

188 **"Underlying public distrust":** The quote comes from Mike Sunnucks in the *Phoenix Business Journal* (September 30, 2008). In an interview by Nathan Gardels on The Huffington Post (September 16, 2008), Joseph Stiglitz, a Nobel Prize–winning economist, claimed: "The fall of Wall Street is for market fundamentalism what the fall of the Berlin Wall was for Communism—it tells the world that this way of economic organization turns out not to be sustainable."

189 **"Heads must roll":** Maureen Dowd, "After W., Le Deluge," *New York Times,* October 19, 2008.

189 **It really was the inequity:** The original capuchin study by Sarah Brosnan and myself was published in 2003, followed by one with more monkeys and stricter controls by Megan van Wolkenten and co-workers (2007). Partially supportive replications were conducted by Grace Fletcher (2008) and Julie Neiworth and co-workers (2009). Brosnan (2008) offers an evolutionary account of primate inequity responses.

190 **observation of a bonobo:** Incident related in *Bonobo* (de Waal, 1997, p. 41) by Sue Savage-Rumbaugh, who believes that bonobos are happiest when everyone gets the same. This species may indeed be more inequity averse than the other apes (Bräuer et al., 2009).

191 **Moe's thirty-ninth birthday:** Described in "The Animal Within," by Amy Argetsinger, *Washington Post,* May 24, 2005. Castration of victims is not unusual among wild male chimps, similar to observation at the Arnhem Zoo mentioned in chapter 2.

191 **"If you want peace":** This quote is being attributed to both American essayist H. L. Mencken (1880–1956) and Pope Paul VI (1897–1978).

192 **fairness norm:** A version of the ultimatum game was tried on chimpanzees by Keith Jensen and co-workers (2007), but since the apes accepted all (including zero) offers, they probably never grasped the game's contingencies (Brosnan, 2008).

192 **"I then had dinner":** From Irene Pepperberg's *Alex & Me* (2008, p. 153).

193 **dogs too may be sensitive to injustice:** Study by Friederike Range et al., (2009). Vilmos Csányi (2005, p. 69) describes the need for dogs to be treated equally: "Dogs keep track of every morsel and every caress and want to have a part in everything. If the master ignores this, they become seriously depressed or aggressive towards the favored individual."

193 **used in a "chimpomat":** Token-exchange studies by John Wolfe (1936).

194 **seeing another's fortune:** This refers to the Adam Smith quote at the start of chapter 1. The study on selfish versus prosocial choices in capuchins was conducted by de Waal et al. (2008). William Harbaugh and co-workers (2007) showed that charity activates reward centers in the human brain.

196 **contrasting Europe and the United States:** Joel Handler (2004). The loss of entrepeneurs was described by Peter Gumbel in "The French Exodus," *Time,* April 5, 2007.

197 **Gini index:** The Gini coefficient measures income distribution from 0 percent (greatest equality) to 100 percent (maximum inequality). With a Gini index of 45 percent, the United States ranks between Uruguay and Cameroon according to the *CIA World Factbook* (2008). Even India (37 percent) and Indonesia (36 percent) have a more equal income distribution, whereas the index is between 25 percent and 35 percent for most European nations. How income inequality harms the economy has been documented by Larry Bartels (2008).

197 **Less egalitarian states:** Utah and New Hampshire (with the most equal income distributions) are healthier than Louisiana and Mississippi (with the most unequal distributions). S. V. Subramanian and Ichiro Kawachi (2003) explored if this effect could be explained away by the racial composition of states, but found the relation to persist if race is taken into account.

197 **Richard Wilkinson:** Quoted from Wilkinson (2006, p. 712). Perhaps the underlying emotions are more basic than suggested, however, because an experiment by Fatemeh Heidary and co-workers (2008) found similar negative health effects in rabbits. For eight weeks, animals were subjected to either food deprivation in isolation (one-third of their normal diet), or the same food deprivation while they could see, hear, and smell well-fed rabbits. The second group showed significantly more signs of stress-related cardiac atrophy.

199 **watched a female baboon assist a male:** Benjamin Beck (1973).

CHAPTER 7: CROOKED TIMBER

201 **"Out of the crooked timber":** Translation of Kant's 1784 statement in German: "Aus so krummem Holze, als woraus der Mensch gemacht ist, kann nichts ganz Gerades gezimmert werden." This "crooked timber" phrase figures in the title of a book by Isaiah Berlin as well as the name of a popular blogsite (Crookedtimber.org).

201 **"We have always known":** A reference to the Great Depression in FDR's Second Inaugural Address (January 20, 1937).

201 **kudzu:** Considered the scourge of the South, because it suffocates anything it overgrows, kudzu is a Japanese vine introduced for erosion control in the 1930s. It can grow one foot per day, and is now beyond control.

202 **glorious New Man:** Quote from Leon Trotsky (1922). See also Steven Pinker (2002) for the communist belief in a flexible human nature.

202 **raised as a girl:** John Colapinto, *As Nature Made Him* (2000), offers the real story of a boy who suffered a botched circumcision, thus becoming a test case for a sexologist who thought that gender is environmentally constructed. The boy's testicles were surgically removed, he received injections of female hormones, and was told he was a girl. This could not undo the prenatal effects of hormones on his brain, however. The child walked like a boy and objected violently to girl clothes and toys. He committed suicide at the age of thirty-eight. Gender identity is now widely recognized as biologically determined.

203 **the gentle, sexy bonobo:** *Our Inner Ape* (de Waal, 2005) discusses human similarities to both of our closest primate relatives: the bonobo and the chimpanzee.

204 **"we don't do body counts":** The destruction of New York's World Trade Center, in 2001, was celebrated across the Muslim world, and the bombing of Baghdad received much flag-waving support in the United States, and was even compared to a symphony by retired Major General Donald Shepperd: "I don't mean to be glib about it, but it really is a symphony that has to be orchestrated by a conductor" (CNN News, March 21, 2003). Rumsfeld's statement on Iraqi dead was given on Fox News (November 2, 2003).

204 **Yosef Lapid:** The justice minister said that the images of Israeli destruction in Gaza reminded him of his family's situation in World War II. Lapid had lost family members in the Holocaust ("Gaza Political Storm Hits Israel," BBC News, May 23, 2004).

205 **"nature is filled with competition":** David Brooks, "Human Nature Redux," *New York Times,* February 17, 2007.

205 **Martin Hoffman:** Hoffman (1981, p. 79).

205 **"mentalize" automatically:** The intentional mental states (e.g., desires, needs, feelings, beliefs, goals, reasons) of others are unobservable constructs deduced from observed behavior. Mentalization helps us make sense of the behavior around us (Allen et al., 2008).

206 **applies equally well to a dog:** Patricia McConnell (2005) interprets canine behavior in emotional terms.

207 **Queen Victoria:** Matt Ridley (2001) describes the first displays of apes at the London Zoo.

207 **Toward the end of a long career:** David Premack (2007) and Jeremy Kagan (2004).

210 **"As long as I live":** J. K. Rowling (2008).

211 **snakes in suits:** Title of a book on psychopathy in business by Paul Babiak and Robert Hare (2006).

212 **a developmental disorder:** James Blair (1995).

212 **polar bear plays with a husky:** Many animals "self-handicap" while playing with younger or weaker partners. A gorilla male, who could kill a juvenile by just leaning a hand on its chest, controls his incredible strength in wrestling and tickling games. Rare play between a polar bear and a tethered sled dog was documented in Canada's Hudson Bay by German photographer Norbert Rosing.

213 **Germany in the previous century:** Robert Waite, *The Psychopathic God: Adolph Hitler* (1977). Hitler has also been diagnosed as a paranoid schizophrenic (Coolidge et al., 2007).

213 **"In those days":** Mark Rowlands (2008, p. 181). A character study of Tertullian concluded that the church father was indeed close to being a psychopath (Nisters, 1950).

214 **open or close the portal:** This is known as the "appraisal" mechanism, that is, the question which cues modulate the empathic response. Major cues for empathy are familiarity and similarity between subject and object (Preston and de Waal, 2002). Further see Frederique de Vignemont and Tania Singer (2006).

214 **the fifth horseman:** Ashley Montagu and Floyd Matson (1983).

214 **hardwired for empathy:** British autism researcher Simon Baron-Cohen (2003) claims that the female brain is specialized in empathizing and the male brain in systematizing. See Carolyne Zahn-Waxler et al. (1992, 2006) for sex differences in childhood and Alan Feingold (1994) for cross-cultural evidence that women are more "tender-minded" and nurturing than men.

215 **"Pity, though it is":** From Bernard de Mandeville's "An Enquiry into the Origin of Moral Virtue" (*Fable of the Bees,* 2nd edition). Mandeville (1670–1733) represents perhaps the closest historical parallel to Ayn Rand's attempt to celebrate egoism as a moral good. The subtitle of Mandeville's satirical fable says it all: "Private Vices, Publick Benefits." Claiming that rapaciousness feeds prosperity, he elevated selfish motives and their economic outcome above other human values.

215 **a more complex picture:** Nancy Eisenberg (2000) and Sara Jaffee and Janet Shibley Hyde (2000) doubt pronounced gender differences in empathy.

216 **Steve Ballmer:** "Ballmer 'vowed to kill Google,'" Ina Fried (CNET News, September 5, 2005).

219 **ancient Greeks:** Ajax, the great warrior, went into a suicidal depression after the Trojan War. Sophocles noted about his insanity: "Now he suffers lonely thoughts. . . ." The U.S. military uses Greek plays as PTSD counseling tools (MSNBC.com, August 14, 2008).

220 **"I am sick and tired of war":** Sherman's and other quotes on killing and warfare come from Dave Grossman's (1995) *On Killing.*

220 **"You saw the ox":** From *The Works of Mencius* (Book I, Part I, Chapter VII).

221 **measure the greatest happiness:** Paul Zak (2005) shows that self-reported happiness correlates positively with generalized trust across nations.

221 **Alan Greenspan:** Quoted by Jim Puzzanghera, *Los Angeles Times*, October 24, 2008.

222 **Smith saw society:** Jonathan Wight (2003).

223 **"dismal science":** About women in economics, see John Kay, "A Little Empathy Would Be Good for Economics," *Financial Times*, June 12, 2003. The term *stakeholder* (which covers the employees, clients, bankers, suppliers, and local community of a business) is increasingly used as counterpoint to *shareholder* or *stockholder*, such as in Edward Freeman's (1984) stakeholder theory.

224 **"We are more compassionate":** From Barack Obama's acceptance speech at the Democratic National Convention in Denver, Colorado (August 28, 2008).

224 **Abraham Lincoln:** In a letter to Joshua Speed (August 24, 1855).

References

CHAPTER 1: BIOLOGY, LEFT AND RIGHT

Bar-Yosef, O. (1986). The walls of Jericho: An alternative interpretation. *Current Anthropology* 27: 157–162.

Behar, D. et al. (2008). The dawn of human matrilineal diversity. *American Journal of Human Genetics* 82: 1130–1140.

Blum, D. (2002). *Love at Goon Park: Harry Harlow and the Science of Affection.* New York: Perseus.

Churchill, W. S. (1991 [orig. 1932]). *Thoughts and Adventures.* New York: Norton.

Darwin, C. (1981 [orig. 1871]). *The Descent of Man, and Selection in Relation to Sex.* Princeton, NJ: Princeton University Press.

de Waal, F. B. M. (1986). The brutal elimination of a rival among captive male chimpanzees. *Ethology & Sociobiology* 7: 237–251.

de Waal, F. B. M. (1997). *Bonobo: The Forgotten Ape,* with photographs by Frans Lanting. Berkeley: University of California Press.

de Waal, F. B. M. (2006). *Primates and Philosophers: How Morality Evolved.* Princeton, NJ: Princeton University Press.

Fry, D. P. (2006). *The Human Potential for Peace: An Anthropological Challenge to Assumptions about War and Violence.* New York: Oxford University Press.

Haidt, J. (2001). The emotional dog and its rational tail: A social intuitionist approach to moral judgment. *Psychological Review* 108: 814–834.

Helliwell, J. F. (2003). How's life? Combining individual and national variables to explain subjective well-being. *Economic Modeling* 20: 331–360.

Hockings, K. J., Anderson, J. R., and Matsuzawa, T. (2006). Road crossing in chimpanzees: A risky business. *Current Biology* 16: 668–670.

Hume, D. (1985 [orig. 1739]). *A Treatise of Human Nature.* Harmondsworth, UK: Penguin.

Kano, T. (1992). *The Last Ape: Pygmy Chimpanzee Behavior and Ecology.* Stanford, CA: Stanford University Press.

Lemov, R. (2005). *World as Laboratory: Experiments with Mice, Mazes, and Men.* New York: Hill & Wang.

Lordkipanidze, D. et al. (2007). Postcranial evidence from early Homo from Dmanisi, Georgia. *Nature* 449: 305–310.

Marshall Thomas, E. (2006). *The Old Way: A Story of the First People.* New York: Sarah Crichton.

Martikainen, P., and Valkonen, T. (1996). Mortality after the death of a spouse: Rates and causes of death in a large Finnish cohort. *American Journal of Public Health* 86: 1087–1093.

Niedenthal, P. M. (2007). Embodying emotion. *Science* 316: 1002–1005.

Poole, J. (1996). *Coming of Age with Elephants: A Memoir.* New York: Hyperion.

Rodseth, L., Wrangham, R. W., Harrigan, A. M., and Smuts, B. B. (1991). The human community as a primate society. *Current Anthropology* 32: 221–254.

Rossiter, C. (1961). *The Federalist Papers.* New York: New American Library.

Rousseau, J-J. (1762 [orig. 1968]). *The Social Contract.* London: Penguin.

Roy, M. M., and Christenfeld, N. J. S. (2004). Do dogs resemble their owners? *Psychological Science* 15: 361–363.

Saffire, W. (1990). The bonding market. *New York Times Magazine* (June 24, 1990).

Smith, A. (1937 [orig. 1759]). *A Theory of Moral Sentiments.* New York: Modern Library.

Smith, A. (1982 [orig. 1776]). *An Inquiry into the Nature and Causes of the Wealth of Nations.* Indianapolis, IN: Liberty Classics.

Thierry, B., and Anderson, J. R. (1986). Adoption in anthropoid primates. *International Journal of Primatology* 7: 191–216.

van Schaik, C. P., and van Noordwijk, M. A. (1985). Evolutionary effect of the absence of felids on the social organization of the macaques on the island of Simeulue. *Folia primatologica* 44: 138–147.

Wiessner, P. (2001). Taking the risk out of risky transactions: A forager's dilemma. In *Risky Business,* F. Salter (Ed.), pp. 21–43. Oxford, UK: Berghahn.

Wrangham, R. W., and Peterson, D. (1996). *Demonic Males: Apes and the Evolution of Human Aggression.* Boston: Houghton Mifflin.

Zajonc, R. B., Adelmann, P. K., Murphy, S. T., and Niedenthal, P. M. (1987). Convergence in the physical appearance of spouses: An implication of the vascular theory of emotional efference. *Motivation & Emotion* 11: 335–346.

CHAPTER 2: THE OTHER DARWINISM

Carnegie, A. (1889). Wealth. *North American Review* 148: 655–657.

Clark, C. (1997). *Misery and Company: Sympathy in Everyday Life.* Chicago: University of Chicago Press.

Dawkins, R. (1976). *The Selfish Gene.* Oxford, UK: Oxford University Press.

de Tocqueville, A. (1969 [orig. 1835]). *Democracy in America,* vol. 1. New York: Anchor.

de Waal, F. B. M. (1996). *Good Natured: The Origins of Right and Wrong in Humans and Other Animals.* Cambridge, MA: Harvard University Press.

de Waal, F. B. M. (1999). Anthropomorphism and anthropodenial: Consistency in our thinking about humans and other animals. *Philosophical Topics* 27: 255–280.

de Waal, F. B. M. (2007 [orig. 1982]). *Chimpanzee Politics.* Baltimore: Johns Hopkins University Press.

Flack, J. C., Krakauer, D. C., and de Waal, F. B. M. (2005). Robustness mechanisms in primate societies: A perturbation study. *Proceedings of the Royal Society London* B 272: 1091–1099.

Ghiselin, M. (1974). *The Economy of Nature and the Evolution of Sex.* Berkeley: University of California Press.

Hofstadter, R. (1992 [orig. 1944]). *Social Darwinism in American Thought.* Boston: Beacon.

Kropotkin, P. (1972 [orig. 1902]). *Mutual Aid: A Factor of Evolution.* New York: New York University Press.

Lott, T. (2005). *Herding Cats: A Life in Politics.* New York: Harper.

Mayr, E. (1961). Cause and effect in biology. *Science* 134: 1501–1506.

McLean, B., and Elkind, P. (2003). *Smartest Guys in the Room: The Amazing Rise and Scandalous Fall of Enron.* New York: Portfolio.

Meston, C. M., and Buss, D. M. (2007). Why humans have sex. *Archives of Sexual Behavior* 36: 477–507.

Midgley, M. (1979). Gene-juggling. *Philosophy* 54: 439–458.

Rand, A. (1992 [orig. 1957]). *Atlas Shrugged.* New York: Dutton.

Ridley, M. (1996). *The Origins of Virtue.* New York: Penguin.

Silk, J. B., Alberts, S. C., and Altmann, J. (2003). Social bonds of female baboons enhance infant survival. *Science* 302: 1231–1234.

Smuts, B. B. (1985). *Sex and Friendship in Baboons.* New York: Aldine.

Solomon, R. C. (2007). Free enterprise, sympathy, and virtue. In *Moral Markets: The Critical Role of Values in the Economy,* P. J. Zak (Ed.), pp. 16–41. Princeton, NJ: Princeton University Press.

Spencer, H. (1864). *Social Statics.* New York: Appleton.

Tinbergen, N. (1963). On aims and methods of ethology. *Zeitschrift für Tierpsychologie* 20: 410–433.

Todes, D. (1989). *Darwin without Malthus: The Struggle for Existence in Russian Evolutionary Thought.* New York: Oxford University Press.

Whybrow, P. C. (2005). *American Mania: When More Is Not Enough.* New York: Norton.

Wright, R. (1994). *The Moral Animal.* New York: Pantheon.

CHAPTER 3: BODIES TALKING TO BODIES

Alexander, R. D. (1986). Ostracism and indirect reciprocity: The reproductive significance of humor. *Ethology & Sociobiology* 7: 253–270.

Anderson, J. R., Myowa-Yamakoshi, M., and Matsuzawa, T. (2004). Contagious yawning in chimpanzees. *Proceedings of the Royal Society of London* B 271: S468–S470.

Aureli, F., Preston, S. D., & de Waal, F. B. M. (1999). Heart rate responses to social interactions in free-moving rhesus macaques: A pilot study. *Journal of Comparative Psychology* 113: 59–65.

Bard, K. A. (2007). Neonatal imitation in chimpanzees tested with two paradigms. *Animal Cognition* 10: 233–242.

Batson, C. D. (1991). *The Altruism Question: Toward a Social-Psychological Answer.* Hillsdale, NJ: Erlbaum.

Boesch, C. (2007). What makes us human (*Homo sapiens*)? The challenge of cognitive cross-species comparison. *Journal of Comparative Psychology* 121: 227–240.

Bonnie, K. E., Horner, V., Whiten, A., and de Waal, F. B. M. (2006). Spread of arbitrary conventions among chimpanzees: A controlled experiment. *Proceedings of the Royal Society of London* B, 274: 367–372.

Chartrand, T. L., and Bargh, J. A. (1999). The chameleon effect: The perception-behavior link and social interaction. *Journal of Personality & Social Psychology* 76: 893–910.

Church, R. M. (1959). Emotional reactions of rats to the pain of others. *Journal of Comparative Physiological Psychology* 52: 132–134.

Cole, J. (2001). Empathy needs a face. *Journals of Consciousness Studies* 8: 51–68.

Darwin, C. (1981 [orig. 1871]). *The Descent of Man, and Selection in Relation to Sex.* Princeton, NJ: Princeton University Press.

Davila Ross, M., Menzler, S., and Zimmermann, E. (2007). Rapid facial mimicry in orangutan play. *Biology Letters* 4: 27–30.

de Gelder, B. (2003). Towards the neurobiology of emotional body language. *Nature Review of Neuroscience* 7: 242–249.

de Waal, F. B. M. (1996). *Good Natured: The Origins of Right and Wrong in Humans and Other Animals.* Cambridge, MA: Harvard University Press.

de Waal, F. B. M. (2001). *The Ape and the Sushi Master.* New York: Basic Books.

de Waal, F. B. M., Boesch, C., Horner, V., and Whiten, A. (2008). Comparing children and apes not so simple. *Science* 319: 569.

Dimberg, U., Thunberg, M., and Elmehed, K. (2000). Unconscious facial reactions to emotional facial expressions. *Psychological Science* 11: 86–89.

Dosa, D. M. (2007). A day in the life of Oscar the cat. *New England Journal of Medicine* 357: 328–329.

Eisenberg, N. (2000). Empathy and sympathy. In *Handbook of Emotion* (2nd ed.), M. Lewis and J. M. Haviland-Jones (Eds.), pp. 677–691. New York: Guilford.

Ferrari P. F., Fogassi, L., Gallese, V., and Rizzolatti, G. (2003). Mirror neurons responding to the observation of ingestive and communicative mouth actions in the monkey ventral premotor cortex. *European Journal of Neuroscience* 17: 1703–1714.

Ferrari, P. F., Visalberghi, E., Paukner, A., Fogassi, L., Ruggiero, A., and Suomi, S. J. (2006). Neonatal imitation in rhesus macaques. *PLoS-Biology* 4: 1501–1508.

Gallese, V. (2005). "Being like me": Self-other identity, mirror neurons, and empathy. In *Perspectives on Imitation,* S. Hurley and N. Chater (Eds.), pp. 101–118. Cambridge, MA: MIT Press.

Gallese, V., Keysers, C., and Rizzolatti, G. (2004). A unifying view of the basis of social cognition. *Trends in Cognitive Science* 8: 396–403.

Geissmann, T., and Orgeldinger, M. (2000). The relationship between duet songs and pair bonds in siamangs, *Hylobates syndactylus. Animal Behaviour* 60: 805–809.

Goodall, J. (1990). *Through a Window.* Boston: Houghton Mifflin.

Hatfield, E., Cacioppo, J. T., and Rapson, R. L. (1994). *Emotional Contagion.* Cambridge, UK: Cambridge University Press.

Haun, D. B. M., and Call, J. (2008). Imitation recognition in great apes. *Current Biology* 18: 288–290.

Herman, L. H. (2002). Vocal, social, and self-imitation by bottlenosed dolphins. In *Imitation in Animals and Artifacts*. K. Dautenhahn and C. L. Nehaniv (Eds.), pp. 63–108. Cambridge, MA: MIT Press.

Herrmann, E., Call, J., Hernàndez-Lloreda, M. V., Hare, B., and Tomasello, M. (2007). Humans have evolved specialized skills of social cognition: The cultural intelligence hypothesis. *Science* 317: 1360–1366.

Hobbes, T. (1991 [orig. 1651]). *Leviathan*. Cambridge, UK: Cambridge University Press.

Hoffman, M. L. (1978). Sex differences in empathy and related behaviors. *Psychological Bulletin* 84: 712–722.

Hopper, L., Spiteri, A., Lambeth, S. P., Schapiro, S. J., Horner, V., and Whiten, A. (2007). Experimental studies of traditions and underlying transmission processes in chimpanzees. *Animal Behaviour* 73: 1021–1032.

Horner, V., and Whiten, A. (2007). Learning from others' mistakes? Limits on understanding a trap-tube task by young chimpanzees and children. *Journal of Comparative Psychology* 121: 12–21.

Horner, V., Whiten, A., Flynn, E., and de Waal, F. B. M. (2006). Faithful replication of foraging techniques along cultural transmission chains by chimpanzees and children. *Proceedings National Academy of Sciences, USA* 103: 13878–13883.

Iacoboni, M. (2005). Neural mechanisms of imitation. *Current Opinion in Neurobiology* 15: 632–637.

Joly-Mascheroni, R. M., Senju, A., and Shepherd, A. J. (2008). Dogs catch human yawns. *Biology Letters* 4: 446–448.

Langford, D. J., et al. (2006). Social modulation of pain as evidence for empathy in mice. *Science* 312: 1967–1970.

Lipps, T. (1903). Einfühlung, innere Nachahmung und Organempfindung. *Archiv für die gesammte Psychologie*, vol. I, part 2. Leipzig: Engelman.

Lucke, J. F., and Batson, C. D. (1980). Response suppression to a distressed conspecific: Are laboratory rats altruistic? *Journal of Experimental Social Psychology* 16: 214–227.

MacLean, P. D. (1985). Brain evolution relating to family, play, and the separation call. *Archives of General Psychiatry* 42: 405–417.

Marshall-Pescini, S., and Whiten, A. (2008). Social learning of nut-cracking behavior in East African sanctuary-living chimpanzees (*Pan troglodytes schweinfurthii*). *Journal of Comparative Psychology* 122: 186–194.

Marshall, J. T., and Sugardjito, J. (1986). Gibbon systematics. In *Comparative Primate Biology*, vol. 1, D. R. Swindler and J. Erwin (Eds.), pp. 137–185. New York: Liss.

Martin, G. B., and Clark, R. D. (1982). Distress crying in neonates: Species and peer specificity. *Developmental Psychology* 18: 3–9.

Masserman, J., Wechkin, M. S., and Terris, W. (1964). Altruistic behavior in rhesus monkeys. *American Journal of Psychiatry* 121: 584–585.

McDougall, W. (1923 [orig. 1908]). *An introduction to Social Psychology.* London: Methuen.

McGrew, W. C. (2004). *The Cultured Chimpanzee: Reflections on Cultural Primatology.* Cambridge, UK: Cambridge University Press.

Meaney, C. A. (2000). In perfect unison. In *The Smile of a Dolphin: Remarkable Accounts of Animal Emotions,* M. Bekoff (Ed.), p. 50. New York: Discovery Books.

Meeren, H. K. M., van Heijnsbergen, C. C. R. J., and de Gelder, B. (2005). Rapid perceptual integration of facial expression and emotional body language. *Proceedings of the National Academy of Sciences, USA* 102: 16518–16523.

Meltzoff, A. N., and Moore, M. K. (1995). A theory of the role of imitation in the emergence of self. In *The Self in Infancy: Theory and Research,* P. Rochat (Ed.), pp. 73–93. Amsterdam: Elsevier.

Merleau-Ponty, M. (1964). *The Primacy of Perception.* Evanston, IL: Northwestern University Press.

Miller, R. E. (1967). Experimental approaches to the physiological and behavioral concomitants of affective communication in rhesus monkeys. In *Social Communication among Primates,* S. A. Altmann (Ed.), pp. 125–134. Chicago: University of Chicago Press.

Miller, R. E., Murphy, J. V., and Mirsky, I. A. (1959). Relevance of facial expression and posture as cues in communication of affect between monkeys. *AMA Archives of General Psychiatry* 1: 480–488.

Moore, B. R. (1992). Avian movement imitation and a new form of mimicry: Tracing the evoluting of a complex form of learning. *Behaviour* 122: 231–263.

Panksepp, J. (1998). *Affective Neuroscience.* New York: Oxford University Press.

Paukner, A., Anderson, J. R., Borelli, E., Visalberghi, E., and Ferrari, P. F. (2005). Macaques recognize when they are being imitated. *Biology Letters* 1: 219–222.

Payne, K. (1998). *Silent Thunder: In the Presence of Elephants.* New York: Penguin.

Platek, S. M., Mohamed, F. B., and Gallup, G. G. (2005). Contagious yawning and the brain. *Cognitive Brain Research* 23: 448–452.

Povinelli, D. J. (2000). *Folk Physics for Apes.* Oxford, UK: Oxford University Press.

Prather, J. F., Peters, S., Nowicki, S., and Mooney, R. (2008). Precise auditory-vocal mirroring in neurons for learned vocal communication. *Nature* 451: 305–310.

Preston, S. D., and de Waal, F. B. M. (2002). Empathy: Its ultimate and proximate bases. *Behavioral & Brain Sciences* 25: 1–72.

Preston, S. D., and Stansfield, R. B. (2008). I know how you feel: Task-irrelevant facial expressions are spontaneously processed at a semantic level. *Cognitive, Affective, & Behavioral Neuroscience* 8: 54–64.

Preston, S. D., Bechara, A., Grabowski, T. J., Damasio, H., and Damasio, A. R. (2007). The neural substrates of cognitive empathy. *Social Neuroscience* 2: 254–275.

Proffitt, D. R. (2006). Embodied perception and the economy of action. *Perspectives on Psychological Science* 1: 110–122.

Provine, R. (2000). *Laughter: A Scientific Investigation*. New York: Viking.

Repp, B. H., and Knoblich, G. (2004). Perceiving action identity: How pianists recognize their own performances. *Psychological Science* 15: 604–609.

Russon, A. E. (1996). Imitation in everyday use: Matching and rehearsal in the spontaneous imitation of rehabilitant orangutans (*Pongo pygmaeus*). In *Reaching into Thought: The Minds of the Great Apes*, A. E. Russon, K. A. Bard, and S. T. Parker (Eds.), pp. 152–176. Cambridge, UK: Cambridge University Press.

Sagi, A., and Hoffman, M. L. (1976). Empathic distress in the newborn. *Developmental Psychology* 12: 175–176.

Schloßberger, M. (2005). *Die Erfahrung des Anderen: Gefühle im menschlichen Miteinander*. Berlin: Akademie Verlag.

Senju, A., Maeda, M., Kikuchi, Y., Hasegawa, T., Tojo, Y., and Osanai, H. (2007). Absence of contagious yawning in children with autism spectrum disorder. *Biology Letters* 3: 706–708.

Singer, T., Seymour, B., O'Doherty, J. P., Stephan, K. E., Dolan, R. J., and Frith, C. D. (2006). Empathic neural responses are modulated by the perceived fairness of others. *Nature* 439: 466–469.

Sisk, J. P. (1993). Saving the world. *First Things* 33: 9–14.

Smith, A. (1976 [orig. 1759]. *A Theory of Moral Sentiments*, D. D. Raphael, A. L. Macfie (Eds.). Oxford, UK: Clarendon Press.

Sonnby-Borgström, M. (2002). Automatic mimicry reactions as related to differences in emotional empathy. *Scandinavian Journal of Psychology* 43: 433–443.

Stürmer, S., Snyder, M., and Omoto, A. M. (2005). Prosocial emotions and helping: The moderating role of group membership. *Journal of Personality & Social Psychology* 88: 532–546.

Taylor, S. (2002). *The Tending Instinct*. New York: Times Books.

Thelen, E., Schoner, G., Scheier, C., and Smith, L. B. (2001). The dynamics of embodiment: A field theory of infant perseverative reaching. *Behavioral & Brain Sciences* 24: 1–86.

Thorndike, E. L. (1898). Animal intelligence: An experimental study of the associative process in animals. *Psychological Review & Monography* 2: 551–553.

Tomasello, M. (1999). *The Cultural Origins of Human Cognition*. Cambridge, MA: Harvard University Press.

van Baaren, R. B., Holland, R. W., Steenaert, B., and van Knippenberg, A. (2003). Mimicry for money: Behavioral consequences of imitation. *Journal of Experimental Social Psychology* 39: 393–398.

van Hooff, J. A. R. A. M. (1972). A comparative approach to the phylogeny of laughter and smiling. In *Non-verbal Communication*, R. Hinde (Ed.), pp. 209–241. Cambridge, UK: Cambridge University Press.

van Schaik, C. P. (2004). *Among Orangutans: Red Apes and the Rise of Human Culture*. Cambridge, MA: Belknap.

Walusinski, O., and Deputte, B. L. (2004). Le bâillement: Phylogenèse, éthologie, nosogénie. *Revue Neurologique* 160: 1011–1021.

Wascher, C. A. F., Isabella Scheiber, I. B. R., and Kotrschal, K. (2008). Heart rate modulation in bystanding geese watching social and non-social events. *Proceedings of the Royal Society of London* B 275: 1653–1659.

Wells, R. S. (2003). Dolphin social complexity: Lessons from long-term study and life history. In *Animal Social Complexity*, F. B. M. de Waal and P. L. Tyack (Eds.), pp. 32–56. Cambridge, MA: Harvard University Press.

Whiten, A., and Ham R. (1992). On the nature and evolution of imitation in the animal kingdom: Reappraisal of a century of research. In *Advances in the Study of Behavior*, vol. 21., J. B. Slater et al. (Eds.), pp. 239–283. New York: Academic Press.

Whiten, A., et al. (1999). Cultures in chimpanzees. *Nature* 399: 682–685.

Whiten, A., Horner, V., and de Waal, F. B. M. (2005). Conformity to cultural norms of tool use in chimpanzees. *Nature* 437: 737–740.

Zahn-Waxler, C., Radke-Yarrow, M., Wagner, E., and Chapman, M. (1992). Development of concern for others. *Developmental Psychology* 28: 126–136.

CHAPTER 4: SOMEONE ELSE'S SHOES

Bailey, M. B. (1986). Every animal is the smartest: Intelligence and the ecological niche. In *Animal Intelligence*, R. Hoage and L. Goldman (Eds.), pp. 105–113. Washington, DC: Smithsonian Institution Press.

Batson, C. D. (1991). *The Altruism Question: Toward a Social-Psychological An-swer.* Hillsdale, NJ: Erlbaum.

Batson, C. D., et al. (1997). Is empathy-induced helping due to self-other merging? *Journal of Personality & Social Psychology* 73: 495–509.

Bradmetz, J., and Schneider, R. (1999). Is Little Red Riding Hood afraid of her grandmother? Cognitive versus emotional response to a false belief. *British Journal of Developmental Psychology* 17: 501–514.

Bugnyar, T., and Heinrich, B. (2005). Ravens, *Corvus corax,* differentiate be-tween knowledgeable and ignorant competitors. *Proceedings of the Royal Society of London* B 272: 1641–1646.

Burkart, J. M., Fehr, E., Efferson, C., and van Schaik, C. P. (2007). Other-regarding preferences in a non-human primate: Common marmosets provision food altruistically. *Proceedings of the National Academy of Sciences, USA* 104: 19762–19766.

Cialdini, R. B., Brown, S. L., Lewis, B. P., Luce, C. L., and Neuberg, S. L. (1997). Reinterpreting the empathy-altruism relationship: When one into one equals oneness. *Journal of Personality & Social Psychology* 73: 481–94.

Cools, A., van Hout, A. J. M., and Nelissen, M. H. J. (2008). Canine reconcil-iation and third-party-initiated postconflict affiliation: Do peacemak-ing social mechanisms in dogs rival those of higher primates? *Ethology* 114: 53–63.

Cordoni, G., and Palagi, E. (2008). Reconciliation in wolves (*Canis lupus*): New evidence for a comparative perspective. *Ethology* 114: 298–308.

Dally, J. M., Emery, N. J., and Clayton, N. S. (2006). Food-caching western scrub-jays keep track of who was watching when. *Science* 312: 1662–1665.

Darley, J. M., and Batson, C. D. (1973). From Jerusalem to Jericho: A study of situational and dispositional variables in helping behavior. *Journal of Personality & Social Psychology* 27: 100–108.

de Gelder, B. (1987). On having a theory of mind. *Cognition* 27: 285–290.

de Waal, F. B. M. (2007 [orig. 1982]). *Chimpanzee Politics: Power and Sex among Apes.* Baltimore: Johns Hopkins University Press.

de Waal, F. B. M. (1989). *Peacemaking among Primates.* Cambridge, MA: Har-vard University Press.

de Waal, F. B. M. (1996). *Good Natured: The Origins of Right and Wrong in Hu-mans and Other Animals.* Cambridge, MA: Harvard University Press.

de Waal, F. B. M. (1997). *Bonobo: The Forgotten Ape.* Berkeley: University of California Press.

de Waal, F. B. M. (2002). *The Ape and the Sushi Master.* New York: Basic Books.

de Waal, F. B. M. (2003). On the possibility of animal empathy. In *Feelings & Emotions: The Amsterdam Symposium*, T. Manstead, N. Frijda, and A. Fischer (Eds.), pp. 379–399. Cambridge, UK: Cambridge University Press.

de Waal, F. B. M. (2008). Putting the altruism back into altruism: The evolution of empathy. *Annual Review of Psychology* 59: 279–300.

de Waal, F. B. M. (in press). The need for a bottom-up account of chimpanzee cognition. In *The Mind of the Chimpanzee: Ecological and Experimental Perspectives*, E. V. Lonsdorf, S. R. Ross, and T. Matsuzawa (Eds). Chicago: University of Chicago Press.

de Waal, F. B. M., Leimgruber, K., and Greenberg, A. R. (2008). Giving is self-rewarding for monkeys. *Proceedings of the National Academy of Sciences, USA* 105: 13685–13689.

de Waal, F. B. M., and van Roosmalen, A. (1979). Reconciliation and consolation among chimpanzees. *Behavioral Ecology & Sociobiology* 5: 55–66.

Flombaum, J. I., and Santos, L. R. (2005). Rhesus monkeys attribute perceptions to others. *Current Biology* 15: 447–452.

Fouts, R., and Mills, T. (1997). *Next of Kin*. New York: Morrow.

Fraser, O., Stahl, D., and Aureli, A. (2008). Stress reduction through consolation in chimpanzees. *Proceedings of the National Academy of Sciences, USA* 105: 8557–8562.

Goodall, J. (1986). *The Chimpanzees of Gombe: Patterns of Behavior*. Cambridge, MA: Belknap.

Goodall, J. (1990). *Through a Window*. Boston: Houghton Mifflin.

Hare, B., Call, J., and Tomasello, M. (2001). Do chimpanzees know what conspecifics know? *Animal Behaviour* 61: 139–151.

Harris, P., Johnson, C. N., Hutton, D., Andrews, G., and Cooke, T. (1989). Young children's theory of mind and emotion. *Cognition & Emotion* 3: 379–400.

Hobson, R. P. (1991). Against the theory of "Theory of Mind." *British Journal of Developmental Psychology* 9: 33–51.

Hoffman, M. L. (1981). Is altruism part of human nature? *Journal of Personality & Social Psychology* 40: 121–137.

Hornstein, H. A. (1991). Empathic distress and altruism: Still inseparable. *Psychological Inquiry* 2: 133–135.

Humphrey, N. (1978). Nature's psychologists. *New Scientist* 78: 900–904.

Jensen, K., Hare, B., Call, J., and Tomasello, M. (2006). What's in it for me? Self-regard precludes altruism and spite in chimpanzees. *Proceedings of the Royal Society of London B* 273: 1013–1021.

Kagan, J. (2000). Human morality is distinctive. *Journal of Consciousness Studies* 7: 46–48.

Krebs, D. L. (1991). Altruism and egoism: A false dichotomy? *Psychological Inquiry* 2: 137–139.

Krensky, S. (2002). *Abe Lincoln and the Muddy Pig.* New York: Simon & Schuster.

Kuroshima, H., Fujita, K. Adachi, I., Iwata, K., and Fuyuki, A. (2003). A capuchin monkey recognizes when people do and do not know the location of food. *Animal Cognition* 6: 283–291.

Ladygina-Kohts, N. N. (2001 [1935]). *Infant Chimpanzee and Human Child: A Classic 1935 Comparative Study of Ape Emotions and Intelligence.* F. B. M. de Waal (Ed.). New York: Oxford University Press.

Lakshminarayanan, V. R., and Santos, L. R. (2008). Capuchin monkeys are sensitive to others' welfare. *Current Biology* 18: R999–R1000.

Lanzetta, J. T., and Englis, B. G. (1989). Expectations of cooperation and competition and their effects on observers' vicarious emotional responses. *Journal of Personality & Social Psychology* 56: 543–554.

Lordkipanidze, D., et al. (2007). Postcranial evidence from early *Homo* from Dmanisi, Georgia. *Nature* 449: 305–310.

MacNeilage, P. F., and Davis, B. L. (2000). On the origin of internal structure of word forms. *Science* 288: 527–531.

McConnell, P. (2005). *For the Love of a Dog.* New York: Ballantine.

Menzel, E. W. (1972). Spontaneous invention of ladders in a group of young chimpanzees. *Folia primatologica* 17: 87–106.

Menzel, E. W. (1974). A group of young chimpanzees in a one-acre field. In *Behavior of Non-human Primates,* vol. 5, A. M. Schrier and F. Stollnitz (Eds.), pp. 83–153. New York: Academic Press.

Menzel, E. W., and Johnson, M. K. (1976). Communication and cognitive organization in humans and other animals. *Annals of the New York Academy of Sciences* 280: 131–142.

Nakamura, M., McGrew, W. C., Marchant, L. F., and Nishida, T. (2000). Social scratch: Another custom in wild chimpanzees? *Primates* 41: 237–248.

O'Connell, S. M. 1995. Empathy in chimpanzees: Evidence for Theory of Mind? *Primates* 36: 397–410.

Ottoni, E. B., and Mannu, M. (2001). Semi–free ranging tufted capuchin monkeys spontaneously use tools to crack open nuts. *International Journal of Primatology* 22: 347–358.

Perner, J., and Ruffman, T. (2005). Infants' insight into the mind: How deep? *Science* 308: 214–216.

Premack, D., and Woodruff, G. (1978). Does the chimpanzee have a theory of mind? *Behavioral and Brain Sciences* 1: 515–526.

Rosenblatt, P. (2006). *Two in a Bed: The Social System of Couple Bed Sharing.* New York: State University of New York Press.

Silk, J. B., et al. (2005). Chimpanzees are indifferent to the welfare of unrelated group members. *Nature* 437: 1357–1359.

Smith, A. (1937 [orig. 1759]). *The Theory of Moral Sentiments.* New York: Modern Library.

Swofford, A. (2003). *Jarhead.* New York: Scribner.

Tomasello, M., and Call, J. (2006). Do chimpanzees know what others see—or only what they are looking at? In *Rational Animals?* S. Hurley and M. Nudds (Eds.), pp. 371–384. Oxford, UK: Oxford University Press.

Virányi, Z., Topál, J., Miklósi, A., and Csányi, V. (2005). A nonverbal test of knowledge attribution: A comparative study on dogs and human infants. *Animal Cognition* 9: 13–26.

Warneken, F., Hare, B., Melis, A. P., Hanus, D., and Tomasello, M. (2007). Spontaneous altruism by chimpanzees and young children. *PLoS Biology* 5: 1414–1420.

Wellman, H. M., Phillips, A. T., and Rodriguez, T. (2000). Young children's understanding of perception, desire, and emotion. *Child Development* 71: 895–912.

Wispé, L. (1991). *The Psychology of Sympathy.* New York: Plenum.

Yerkes, R. M. (1925). *Almost Human.* New York: Century.

Zahn-Waxler, C., Hollenbeck, B., and Radke-Yarrow, M. (1984). The origins of empathy and altruism. In *Advances in Animal Welfare Science,* M. W. Fox and L. D. Mickley (Eds.), pp. 21–39. Washington, DC: Humane Society of the United States.

Zillmann, D., and Cantor, J. R. (1977). Affective responses to the emotions of a protagonist. *Journal of Experimental Social Psychology* 13: 155–165.

CHAPTER 5: THE ELEPHANT IN THE ROOM

Anderson, J. R., and Gallup, G. G., Jr. (1999). Self-recognition in nonhuman primates: Past and future challenges. In *Animal Models of Human Emotion and Cognition,* M. Haug and R. E. Whalen (Eds.), pp. 175–194. Washington, DC: APA.

Balfour, D., and Balfour, S. (1998). *African Elephants, A Celebration of Majesty.* New York: Abbeville Press.

Bates, L. A., et al. (2008). Do elephants show empathy? *Journal of Consciousness Studies* 15: 204–225.

Bekoff, M. (2001). Observations of scent-marking and discriminating self from others by a domestic dog: Tales of displaced yellow snow. *Behavioural Processes* 55: 75–79.

Bekoff, M., and Sherman, P. W. (2003). Reflections on animal selves. *Trends in Ecology and Evolution* 19: 176–180.

Bischof-Köhler, D. (1988). Über den Zusammenhang von Empathie und der Fähigkeit sich im Spiegel zu erkennen. *Schweizerische Zeitschrift für Psychologie* 47: 147–159.

Bischof-Köhler, D. (1991). The development of empathy in infants. In *Infant Development: Perspectives from German-Speaking Countries,* M. Lamb and M. Keller (Eds.), pp. 245–273. Hillsdale, NJ: Erlbaum.

Bonnie, K. E., and de Waal, F. B. M. (2004). Primate social reciprocity and the origin of gratitude. In *The Psychology of Gratitude,* R. A. Emmons and M. E. McCullough (Eds.), pp. 213–229. Oxford, UK: Oxford University Press.

Butterworth, G., and Grover, L. (1988). The origins of referential communication in human infancy. In *Thought without Language,* L. Weiskrantz (Ed.), pp. 5–24. Oxford, UK: Clarendon.

Caldwell, M. C., and Caldwell, D. K. (1966). Epimeletic (care-giving) behavior in Cetacea. In *Whales, Dolphins, and Porpoises,* K. S. Norris (Ed.), pp. 755–789. Berkeley: University of California Press.

Carpenter, C. R. (1934). A field study of the behavior and social relations of howling monkeys. *Comparative Psychology Monographs* 10: 1–168.

Cenami Spada, E., Aureli, F., Verbeek, P., and de Waal, F. B. M. (1995). The self as reference point: Can animals do without it? In *The Self in Infancy: Theory and Research,* P. Rochat (Ed.), pp. 193–215. Amsterdam: Elsevier.

Cheney, D. L., and Seyfarth, R. M. (2008). *Baboon Metaphysics: The Evolution of a Social Mind.* Chicago: University of Chicago Press.

Connor, R. C., and Norris, K. S. (1982). Are dolphins reciprocal altruists? *American Naturalist* 119: 358–372.

de Waal, F. B. M. (2007 [orig. 1982]). *Chimpanzee Politics: Power and Sex among Apes.* Baltimore, MD: Johns Hopkins University Press.

de Waal, F. B. M. (1988). The communicative repertoire of captive bonobos (*Pan paniscus*), compared to that of chimpanzees. *Behaviour* 106: 183–251.

de Waal, F. B. M. (1997). *Bonobo: The Forgotten Ape.* Berkeley: University of California Press.

de Waal, F. B. M. (2001). *The Ape and the Sushi Master.* New York: Basic Books.

de Waal, F. B. M. (2008). Putting the altruism back into altruism: The evolution of empathy. *Annual Review of Psychology* 59: 279–300.

de Waal, F. B. M., and Aureli, F. (1996). Consolation, reconciliation, and a possible cognitive difference between macaque and chimpanzee. In *Reaching into Thought: The Minds of the Great Apes,* A. E. Russon, K. A. Bard, and S. T. Parker (Eds.), pp. 80–110. Cambridge, UK: Cambridge University Press.

de Waal, F. B. M., Dindo, M., Freeman, C. A., and Hall, M. (2005). The monkey in the mirror: Hardly a stranger. *Proceedings of the National Academy of Sciences, USA* 102: 11140–11147.

Decety, J., and Chaminade, T. (2003). When the self represents the other: A new cognitive neuroscience view on psychological identification. *Consciousness & Cognition* 12: 577–596.

Decety, J., and Grèzes, J. (2006). The power of simulation: Imagining one's own and other's behavior. *Brain Research* 1079: 4–14.

Douglas-Hamilton, I., Bhalla, S., Wittemyer, G., and Vollrath, F. (2006). Behavioural reactions of elephants towards a dying and deceased matriarch. *Applied Animal Behaviour Science* 100: 87–102.

Emery, N. J., and Clayton, N. S. (2001). Effects of experience and social context on prospective caching strategies by scrub jays. *Nature* 414: 443–446.

Emery, N. J., and Clayton, N. S. (2004). The mentality of crows: Convergent evolution of intelligence in corvids and apes. *Science* 306: 1903–1907.

Engh, A. L., et al. (2005). Behavioural and hormonal responses to predation in female chacma baboons. *Proceedings of the Royal Society of London B* 273: 707–712.

Epstein, R., Lanza, R. P., and Skinner, B. F. (1981). "Self-awareness" in the pigeon. *Science* 212: 695–696.

Gallup, G. G. Jr. (1970). Chimpanzees: Self-recognition. *Science* 167: 86–87.

Gallup, G. G. Jr. (1983). Toward a comparative psychology of mind. In *Animal Cognition and Behavior,* R. L. Mellgren (Ed.), pp. 473–510. New York: North-Holland.

Goleman, D. (2006). *Social Intelligence: The New Science of Human Relationships.* New York: Bantam Books.

Gould, S. J. (1977). *Ontogeny and Phylogeny.* Cambridge, MA: Harvard University Press.

Hakeem, A. Y., Sherwood, C. C., Bonar, C. J., Butti, C., Hof, P. R., and Allman, J. M. (2009). Von Economo Neurons in the elephant brain. *Anatomical Record* 292: 242–248.

Johnson, D. B. (1992). Altruistic behavior and the development of the self in infants. *Merrill-Palmer Quarterly of Behavior & Development* 28: 379–388.

Jorgensen, M. J., Hopkins, W. D., and Suomi, S. J. (1995). Using a computerized testing system to investigate the preconceptual self in nonhuman primates and humans. In *The Self in Infancy: Theory and Research*, P. Rochat (Ed.), pp. 243–256. Amsterdam: Elsevier.

Ladygina-Kohts, N. N. (2001 [1935]). *Infant Chimpanzee and Human Child: A Classic 1935 Comparative Study of Ape Emotions and Intelligence.* F. B. M. de Waal (Ed.). New York: Oxford University Press.

Krause, M. A. (1997). Comparative perspectives on pointing and joint attention in children and apes. *International Journal of Comparative Psychology* 10: 137–157.

Kummer, H. (1968). *Social Organization of Hamadryas Baboons.* Chicago: University of Chicago Press.

Leavens, D. A., and Hopkins, W. D. (1998). Intentional communication by chimpanzees: A cross-sectional study of the use of referential gestures. *Developmental Psychology* 34: 813–822.

Leavens, D. A., and Hopkins, W. D. (1999). The whole-hand point: The structure and function of pointing from a comparative perspective. *Journal of Comparative Psychology* 113: 417–425.

Lewis, M., and Ramsay, D. (2004). Development of self-recognition, personal pronoun use, and pretend play during the 2nd year. *Child Development* 75: 1821–1831.

Manger, P. R. (2006). An examination of cetacean brain structure with a novel hypothesis correlating thermogenesis to the evolution of a big brain. *Biological Review* 81: 293–338.

Marais, E. N. (1939). *My Friends the Baboons.* New York: McBride.

Marino, L. (1998). A comparison of encephalization between odontocete cetaceans and anthropoid primates. *Brain, Behavior, and Evolution* 51: 230–238.

Marino, L., et al. (2007). Cetaceans have complex brains for complex cognition. *PLoS-Biology* 5: e139.

Medicus, G. (1992). The inapplicability of the biogenetic rule to behavioral development. *Human Development* 35: 1–8.

Menzel, C. R. (1999). Unprompted recall and reporting of hidden objects by a chimpanzee (*Pan troglodytes*) after extended delays. *Journal of Comparative Psychology* 113: 426–434.

Menzel, E. W. (1973). Leadership and communication in young chimpanzees. In *Precultural Primate Behavior*, E. W. Menzel (Ed.). Basel: Karger.

Menzel, E. W. (1979). Communication of object-locations in a group of young chimpanzees. In *The Great Apes,* D. A. Hamburg and E. R. McCown (Eds.), pp. 359–371. Menlo Park, CA: Benjamin Cummings.

Moss, C. (1988). *Elephant Memories: Thirteen Years in the Life of an Elephant Family.* New York: Fawcett Columbine.

Nimchinsky, E. A., et al. (1999). A neuronal morphologic type unique to humans and great apes. *Proceedings of the National Academy of Sciences, USA* 96: 5268–5273.

Plotnik, J., de Waal, F. B. M., and Reiss, D. (2006). Self-recognition in an Asian elephant. *Proceedings of the National Academy of Sciences, USA* 103: 17053–17057.

Poole, J. (1996). *Coming of Age with Elephants: A Memoir.* New York: Hyperion.

Povinelli, D. J. (1989). Failure to find self-recognition in Asian elephants (*Elephas maximus*) in contrast to their use of mirror cues to discover hidden food. *Journal of Comparative Psychology* 103: 122–131.

Povinelli, D. J., and Cant, J. G. H. (1995). Arboreal clambering and the evolution of self-conception. *Quarterly Review of Biology* 70: 393–421.

Preston, S. D., and de Waal, F. B. M. (2002). Empathy: Its ultimate and proximate bases. *Behavioral & Brain Sciences* 25: 1–72.

Reiss, D., and Marino, L. (2001). Mirror self-recognition in the bottlenose dolphin: A case of cognitive convergence. *Proceedings of the National Academy of Sciences, USA* 98: 5937–5942.

Rochat, P. (2003). Five levels of self-awareness as they unfold early in life. *Consciousness & Cognition* 12: 717–731.

Sapolsky, R. M. (2001). *A Primate's Memoir: A Neuroscientist's Unconventional Life among the Baboons.* New York: Scribner.

Schino, G., Geminiani, S., Rosati, L., and Aureli, F. (2004). Behavioral and emotional response of Japanese macaque mothers after their offspring receive an aggression. *Journal of Comparative Psychology* 118: 340–346.

Seed, A. M., Clayton, N. S., and Emery, N. J. (2007). Postconflict third-party affiliation in rooks. *Current Biology* 17: 152–158.

Seeley, W. W., et al. (2006). Early frontotemporal dementia targets neurons unique to apes and humans. *Annals of Neurology* 60: 660–667.

Semendeferi, K., Lu, A., Schenker, N., and Damasio, H. (2002). Humans and great apes share a large frontal cortex. *Nature Neuroscience* 5: 272–276.

Siebenaler, J. B., and Caldwell, D. K. (1956). Cooperation among adult dolphins. *Journal of Mammalogy* 37: 126–128.

Smuts, B. B. (1999 [orig. 1985]). *Sex and Friendship in Baboons.* Cambridge, MA: Harvard University Press.

Thompson, R. K. R., and Contie, C. L. (1994). Further reflections on mirror usage by pigeons: Lessons from Winnie-the-Pooh and Pinocchio too. In *Self-Awareness in Animals and Humans*, S. T. Parker et al. (Eds.), pp. 392–409. Cambridge, UK: Cambridge University Press.

Toda, K., and Watanabe, S. (2008). Discrimination of moving video images of self by pigeons (*Columba livia*). *Animal Cognition* 11: 699–705.

Tomasello, M., Carpenter, M., and Liszkowski, U. (2007). A new look at infant pointing. *Child Development* 78: 705–722.

Trivers, R. L. (1971). The evolution of reciprocal altruism. *Quarterly Review of Biology* 46: 35–57.

Veà, J. J., and Sabater-Pi, J. (1998). Spontaneous pointing behaviour in the wild pygmy chimpanzee. *Folia primatologica* 69: 289–290.

Verbeek, P., and de Waal, F. B. M. (1997). Postconflict behavior in captive brown capuchins in the presence and absence of attractive food. *International Journal of Primatology* 18: 703–725.

von Hoesch, W. (1961). Über Ziegenhütende Bärenpaviane. *Zeitschrft für Tierpsychologie* 18: 297–301.

Zahn-Waxler, C., Radke-Yarrow, M., Wagner, E., and Chapman, M. (1992). Development of concern for others. *Developmental Psychology* 28: 126–136.

CHAPTER 6: FAIR IS FAIR

Ahn, T. K., Ostrom, E., Schmidt, D., and Walker, J. (2003). Trust in two-person games: Game structures and linkages. In *Trust and Reciprocity*, E. Ostrom and J. Walker (Eds.), pp. 323–351. New York: Russell Sage.

Alvard, M. (2004). The Ultimatum Game, fairness, and cooperation among big game hunters. In *Foundations of Human Sociality: Ethnography and Experiments in 15 Small-scale Societies*, J. Henrich et al. (Eds.), pp. 413–435. London: Oxford University Press.

Amati, S., Babweteera, and Wittig, R. M. (2008). Snare removal by a chimpanzee of the Sonso community, Budongo Forest (Uganda). *Pan Africa News* 15: 6–8.

Bartels, L. M. (2008). *The Political Economy of the New Gilded Age*. Princeton, NJ: Princeton University Press.

Beck, B. B. (1973). Cooperative tool use by captive Hamadryas baboons. *Science* 182: 594–597.

Bellugi, U., Lichtenberger, L., Jones, W., Lai, Z., and St. George, M. (2000). The neurocognitive profile of Williams Syndrome: A complex pattern of strengths and weaknesses. *Journal of Cognitive Neuroscience* 12: 7–29.

Boehm, C. (1993). Egalitarian behavior and reverse dominance hierarchy. *Current Anthropology* 34: 227–254.

Boesch, C., and Boesch-Achermann, H. (2000). *The Chimpanzees of the Taï Forest.* Oxford, UK: Oxford University Press.

Boyd, R. (2006). The puzzle of human sociality. *Science* 314: 1555–1556.

Bräuer, J., Call, J., and Tomasello, M. (2009). Are apes inequity averse? New data on the token-exchange paradigm. *American Journal of Primatology* 71: 175–181.

Brosnan, S. F. (2008). Responses to inequity in non-human primates. In *Neuroeconomics: Decision Making and the Brain,* P. W. Glimcher et al. (Eds.), pp. 283–300. New York: Academic Press.

Brosnan, S. F., and de Waal, F. B. M. (2003). Monkeys reject unequal pay. *Nature* 425: 297–299.

Brosnan, S. F., and de Waal, F. B. M. (2004). Socially learned preferences for differentially rewarded tokens in the brown capuchin monkey. *Journal of Comparative Psychology* 118: 133–139.

Brosnan, S. F., Schiff, H., and de Waal, F. B. M. (2005). Tolerance for inequity increases with social closeness in chimpanzees. *Proceedings of the Royal Society of London* B 272: 253–258.

Bshary, R., and Würth, M. (2001). Cleaner fish *Labroides dimidiatus* manipulate client reef fish by providing tactile stimulation. *Proceedings of the Royal Society of London* B 268: 1495–1501.

Burnham, T. C., and Johnson, D. D. P. (2005). The biological and evolutionary logic of human cooperation. *Analyse & Kritik* 27: 113–135.

Chase, I. (1988). The vacancy chain process: A new mechanism of resource distribution in animals with application to hermit crabs. *Animal Behaviour* 36: 1265–1274.

Clark, M. S., and Grote N. K. (2003). Close relationships. In *Handbook of Psychology: Personality and Social Psychology,* T. Millon and M. J. Lerner (Eds.), pp. 447–461. New York: John Wiley.

Csányi (2005). *If Dogs Could Talk: Exploring the Canine Mind.* New York: North Point.

Dana, J. D., Kuang, J., and Weber, R. A. (2004). Exploiting moral wriggle room: Behavior inconsistent with a preference for fair outcomes. Available at *Social Science Research Network,* abstract 400900.

de Waal, F. B. M. (2007 [orig. 1982]). *Chimpanzee Politics.* Baltimore: Johns Hopkins University Press.

de Waal, F. B. M. (1989). Food sharing and reciprocal obligations in chimpanzees. *Journal of Human Evolution* 18: 433–459.

de Waal, F. B. M. (1996). *Good Natured: The Origins of Right and Wrong in Humans and Other Animals.* Cambridge, MA: Harvard University Press.

de Waal, F. B. M. (1997). *Bonobo: The Forgotten Ape,* with photographs by Frans Lanting. Berkeley: University of California Press.

de Waal, F. B. M. (1997). The chimpanzee's service economy: Food for grooming. *Evolution of Human Behavior* 18: 375–86.

de Waal, F. B. M. (2000). Attitudinal reciprocity in food sharing among brown capuchins. *Animal Behaviour* 60: 253–261.

de Waal, F. B. M., and Berger, M. L. (2000). Payment for labour in monkeys. *Nature* 404: 563.

de Waal, F. B. M., Leimgruber, K., and Greenberg, A. R. (2008). Giving is self-rewarding for monkeys. *Proceedings of the National Academy of Sciences, USA* 105: 13685–13689.

de Waal, F. B. M., and Luttrell, L. M. (1988). Mechanisms of social reciprocity in three primate species: Symmetrical relationship characteristics or cognition? *Ethology & Sociobiology* 9: 101–118.

Fehr, E., and Fischbacher, U. (2003), The nature of altruism. *Nature* 425, 785–791.

Fehr, E., and Schmidt, K. M. (1999). A theory of fairness, competition, and cooperation. *Quarterly Journal of Economics* 114: 817–868.

Fletcher, G. E. (2008). Attending to the outcome of others: Disadvantageous inequity aversion in male capuchin monkeys. *American Journal of Primatology* 70: 901–905.

Frank, R. H. (1988). *Passions Within Reason.* New York: Norton.

Frank, R. H., and Cook, P. J. (1995). *Winner-Take-All Society.* New York: Free Press.

Freud, S. (1950 [orig. 1913]). *Totem and Taboo: Some Points of Agreement between the Mental Lives of Savages and Neurotics.* New York: Norton.

Gintis, H., Bowles, S., Boyd, R., and Fehr, E. (2005). *Moral Sentiments and Material Interests.* Cambridge, MA: MIT Press.

Güth, W., Schmittberger, R., and Schwarze, B. (1982). An experimental analysis of ultimatum bargaining. *Journal of Economic Behavior & Organization* 3: 367–388.

Handler, J. F. (2004). *Social Citizenship and Workfare in the United States and Western Europe: The Paradox of Inclusion.* Cambridge, UK: Cambridge University Press.

Harbaugh, W. T., Mayr, U., and Burghart, D. R. (2007). Neural responses to taxation and voluntary giving reveal motives for charitable donations. *Science* 326:1622–1625.

Heidary, F., et al. (2008). Food inequality negatively impacts cardiac health in rabbits. *PLoS ONE* 3(11): e3705. doi:10.1371/journal.pone. 0003705.

Henrich, J., Boyd, R., Bowles, S., Camerer, C., Gintis, H., McElreath, R., and Fehr, E. (2001). In search of *Homo economicus*: Experiments in 15 small-scale societies. *American Economic Review* 91: 73–79.

Henzi, S. P. and Barrett, L. (2002). Infants as a commodity in a baboon market. *Animal Behaviour* 63: 915–921.

Hobbes, T. (2004 [orig. 1651]). *De Cive*. Whitefish, MT: Kessinger.

Hockings, K. J., et al. (2007). Chimpanzees share forbidden fruit. *PLoS ONE* 9: e886.

Jensen, K., Call, J., and Tomasello, M. (2007). Chimpanzees are rational maximizers in an Ultimatum Game. *Science* 318: 107–109.

Kahneman, D., Knetsch, J., and Thaler, R. (1986). Fairness and the assumptions of economics. *Journal of Business* 59: 285–300.

Knutson, B., Wimmer, G. E., Kuhnen, C. M., and Winkielman, P. (2008). Nucleus accumbens activation mediates the influence of reward cues on financial risk taking. *NeuroReport* 19: 509–513.

Koyama, N. F., Caws, C., and Aureli, F. (2006). Interchange of grooming and agonistic support in chimpanzees. *International Journal of Primatology* 27: 1293–1309.

Kropotkin, P. (1906). *The Conquest of Bread*. New York: Putnam.

Kummer, H. (1991). Evolutionary transformations of possessive behavior. *Journal of Social Behavior and Personality* 6: 75–83.

Langergraber, K. E., Mitani, J. C., and Vigilant, L. (2007). The limited impact of kinship on cooperation in wild chimpanzees. *Proceedings of the National Academy of Sciences, USA* 104: 7786–7790.

Neiworth, J. J., Johnson, E. T., Whillock, K., Greenberg, J., and Brown, V. (2009). Is a sense of inequity an ancestral primate trait? Testing social inequity in cotton top tamarins (*Saguinus oedipus*). *Journal of Comparative Psychology* 123: 10–17.

Némirovsky, I. (2006). *Suite Française*. New York: Knopf.

Nishida, T., Hasegawa, T., Hayaki, H., Takahata, Y., and Uehara, S. (1992). Meat-sharing as a coalition strategy by an alpha male chimpanzee? In *Topics of Primatology*, T. Nishida (Ed.), pp. 159–174. Tokyo: Tokyo Press.

Noë, R., and Hammerstein, P. (1994). Biological markets: Supply and demand determine the effect of partner choice in cooperation, mutualism and mating. *Behavioral Ecology & Sociobiology* 35: 1–11.

Pepperberg, I. M. (2008). *Alex & Me*. New York: Collins.

Perry, S. (2008). *Manipulative Monkeys: The Capuchins of Lomas Barbudal*. Cambridge, MA: Harvard University Press.

Perry, S., and Rose, L. (1994). Begging and transfer of coati meat by white-faced capuchin monkeys, *Cebus capucinus*. *Primates* 35: 409–415.

Perry, S., et al. (2003). Social conventions in wild white-faced capuchin monkeys: Evidence for traditions in a neotropical primate. *Current Anthropology* 44: 241–268.

Range, F., Horn, L., Viranyi, Z., and Huber, L. (2009). The absence of reward induces inequity aversion in dogs. *Proceedings of the National Academy of Sciences, USA* 106: 340–345.

Rose, L. (1997). Vertebrate predation and food-sharing in Cebus and Pan. *International Journal of Primatology* 18: 727–765.

Rosenau, P. V. (2006). Is economic theory wrong about human nature? *Journal of Economic and Social Policy* 10: 16–78.

Sanfey, A. G., Rilling, J. K., Aronson, J. A., Nystrom, L. E., and Cohen, J. D. (2003). The neural basis of economic decision-making in the ultimatum game. *Science* 300: 1755–1758.

Shermer, M. (2008). *The Mind of the Market*. New York: Times Books.

Skyrms, B. (2004). *The Stag Hunt and the Evolution of Social Structure*. Cambridge, UK: Cambridge University Press.

Smith, A. (1982 [orig. 1776]). *An Inquiry into the Nature and Causes of the Wealth of Nations*. Indianapolis, IN: Liberty Classics.

Subramanian, S. V., and Kawachi, I. (2003). The association between state income inequality and worse health is not confounded by race. *International Journal of Epidemiology* 32: 1022–1028.

Trivers, R. (2004). Mutual benefits at all levels of life. *Science* 304: 964–965.

van Wolkenten, M., Brosnan, S. F., and de Waal, F. B. M. (2007). Inequity responses of monkeys modified by effort. *Proceedings of the National Academy of Sciences, USA* 104: 18854–18859.

Visalberghi, V., and Anderson, J. (2008). Fair game for chimpanzees. *Science* 319: 283–284.

Wilkinson, G. S. (1988). Reciprocal altruism in bats and other mammals. *Ethology & Sociobiology* 9: 85–100.

Wilkinson, R. G. (2006). The impact of inequality. *Social Research* 73: 711–732.

Wolfe, J. B. (1936). Effectiveness of token-rewards for chimpanzees. *Comparative Psychology Monographs* 12 (5): 1–72.

Zak, P. (2008). *Moral Markets*. Princeton, NJ: Princeton University Press.

CHAPTER 7: CROOKED TIMBER

Allen, J. G., Fonagy, P., and Bateman, A. W. (2008). *Mentalizing in Clinical Practice.* Arlington, VA: American Psychiatric Publishing.

Babiak, P., and Hare, R. D. (2006). *Snakes in Suits: When Psychopaths Go to Work.* New York: Collins.

Baron-Cohen, S. (2003). *The Essential Difference: The Truth About the Male and Female Brain.* New York: Basic Books.

Blair, R. J. R. (1995). A cognitive developmental approach to morality: Investigating the psychopath. *Cognition* 57: 1–29.

Colapinto, J. (2000). *As Nature Made Him: The Boy Who Was Raised as a Girl.* New York: HarperCollins.

Coolidge, F. L., Davis, F. L., and Segal, D. L. (2007). Understanding madmen: A DSM-IV assessment of Adolf Hitler. *Individual Differences Research* 5: 30–43.

de Mandeville, B. (1966 [1714]). *The Fable of the Bees: or Private Vices, Publick Benefits,* vol. I. London: Oxford University Press.

de Vignemont, F., and Singer, T. (2006). The empathic brain: How, when and why? *Trends in Cognitive Sciences* 10: 435–441.

Eisenberg, N. (2000). Empathy and sympathy. In *Handbook of Emotion,* M. Lewis and J. M. Haviland-Jones (Eds.), pp. 677–691. New York: Guilford Press.

Feingold, A. (1994). Gender differences in personality: A meta-analysis. *Psychological Bulletin* 116: 429–456.

Freeman, R. E. (1984). *Strategic Management: A Stakeholder Approach.* Boston: Pitman.

Grossman, D. (1995). *On Killing: The Psychological Cost of Learning to Kill in War and Society.* New York: Back Bay Books.

Hoffman, M. L. (1981). Perspectives on the difference between understanding people and understanding things: The role of affect. In *Social Cognitive Development,* J. H. Flavell and L. Ross (Eds.), pp. 67–81. Cambridge, UK: Cambridge University Press.

Jaffee, S., and Hyde, J. S. (2000). Gender differences in moral orientation: A meta-analysis. *Psychological Bulletin* 126: 703–726.

Kagan, J. (2004). The uniquely human in human nature. *Daedalus* 133 (4): 77–88.

Kant, I. (1784). Idee zu einer allgemeinen Geschichte in weltbürgerlicher Absicht. *Berlinische Monatsschrift,* November: 385–411.

McConnell, P. B. (2005). *For the Love of a Dog: Understanding Emotions in You and Your Best Friend.* New York: Ballantine.

Mencius (1895 [orig. fourth century B.C.]). *The Works of Mencius.* Translation: J. Legge. Oxford, UK: Clarendon.

Montagu, A., and Matson, F. (1983). *The Dehumanization of Man.* New York: McGraw-Hill.

Nisters, B. (1950). Tertullian: Seine Persönlichkeit und sein Schicksal. *Münsterische Beiträge zur Theologie* 25.

Pinker, S. (2002). *The Blank Slate: The Modern Denial of Human Nature.* New York: Viking.

Premack, D. (2007). Human and animal cognition: Continuity and discontinuity. *Proceedings of the National Academy of Sciences, USA* 104: 13861–13867.

Preston, S. D., and de Waal F. B. M. (2002). Empathy: Its ultimate and proximate bases. *Behavioral & Brain Sciences* 25: 1–72.

Ridley, M. (2001). Re-reading Darwin. *Prospect* 66: 74–76.

Rowlands, M. (2008). *The Philosopher and the Wolf.* London: Granta.

Rowling, J. K. (2008). Magic for Muggles. *Greater Good* V (1): 40.

Sapolsky, R. M., and Share, L. J. (1998). Darting terrestrial primates in the wild: A primer. *American Journal of Primatology* 44: 155–167.

Singer, T., Seymour, B., O'Doherty, J. P., Stephan, K. E., Dolan, R. J., and Frith, C. D. (2006). Empathic neural responses are modulated by the perceived fairness of others. *Nature* 439: 466–469.

Trotsky, L. (1922). The tasks of communist education. *Communist Review* 4 (7).

Waite, R. (1977). *The Psychopathic God: Adolph Hitler.* New York: Basic Books.

Wight, J. B. (2003). Teaching the ethical foundations of economics. *Chronicle of Higher Education* (Aug. 15, 2003).

Zahn-Waxler, C., Crick, N., Shirtcliff, E. A., and Woods, K. (2006). The origins and development of psychopathology in females and males. In *Developmental Psychopathology,* 2nd ed., vol. I, D. Cicchetti and D. J. Cohen (Eds.), pp. 76–138. New York: John Wiley.

Zahn-Waxler, C., Radke-Yarrow, M., Wagner, E., and Chapman, M. (1992). Development of concern for others. *Developmental Psychology* 28: 126–36.

Zak, P. J. (2005). The neuroeconomics of trust. Available at *Social Science Research Network,* abstract 764944.

Index

Acknowledgments

For about a decade, I have gathered information on the role of empathy and trust in society—both human and animal—for *The Age of Empathy*. This book owes much to many people, especially my changing team of up to twenty students, technicians, and scientists at the Living Links Center, which is part of the Yerkes National Primate Research Center at Emory University in Atlanta, Georgia. Let me thank by name those co-workers, colleagues, and friends who have provided feedback on parts of the manuscript or offered observations, ideas, and quotes. I am grateful to John Allman, Filippo Aureli, Christophe Boesch, Peter Bos, Sarah Brosnan, Devyn Carter, Marietta Dindo, Pier Francesco Ferrari, Jessica Flack, Robert Frank, Amy Fultz, Beatrice de Gelder, Milton Harris, Yuko Hattori, Victoria Horner, Scott Lilienfeld, Charles Menzel, Alison Nash, Mathias Osvath, Susan Perry, Ing-Marie Persson, Diana Reiss, Colleen Schaffner, Anindya Sinha, Susan Stanich, Benjamin de Waal, Polly Wiesner, and Tiffany Young.

I further thank Toshisada Nishida for inviting me to his camp at the Mahale Mountains in Tanzania, Joshua Plotnik and Richard Lair for their hospitality when I came to see elephants in Thailand, Maria Butovskaya for arranging a behind-the-scenes tour of the state Darwin Museum in Moscow, Emil Menzel for graciously agreeing to an interview about his pioneering ideas, the late Wim Suermondt for teaching me to draw, and Stephanie Preston for helping me develop core ideas about the way empathy works. Our research is made possible by funding from the National Science Foundation, the National Institutes of Health, Emory University, and private donations. I further thank my agent, Michelle Tessler, for her continued support, and John Glusman of Harmony Books for his encouragement and critical reading of the entire text.

My first reader, as always, has been my wife, Catherine Marin, who makes sure the text is clear and readable, and who never fails to brighten my life.

ABOUT THE AUTHOR

FRANS DE WAAL received a Ph.D. in biology from the University of Utrecht, in his native Holland, after which he moved to the United States, in 1981. His first book, *Chimpanzee Politics*, compared the schmoozing and scheming of chimpanzees involved in power struggles with that of human politicians. Ever since, de Waal has drawn parallels between primate and human behavior, from aggression to morality and culture. His popular books—translated into more than fifteen languages—have made him one of the world's best-known primatologists.

With his discovery of reconciliation in primates, de Waal founded the field of animal conflict resolution studies. He received the 1989 *Los Angeles Times* Book Award for *Peacemaking among Primates*. His scientific articles have been published in journals such as *Science, Nature, Scientific American*, and outlets specializing in animal behavior. De Waal is C. H. Candler professor in the Psychology Department of Emory University and director of the Living Links Center at the Yerkes National Primate Research Center in Atlanta. He has been elected to the (U.S.) National Academy of Sciences and the Royal Dutch Academy of Sciences. In 2007, *Time* selected him one of the World's 100 Most Influential People.

With his wife, Catherine, and their cats, de Waal lives in a forested area near Stone Mountain, Georgia.

For more on the author and his book, see:
www.emory.edu/LIVING_LINKS/Empathy